T0130754

Methods for Ecological Research
on Terrestrial Small Mammals

Methods for Ecological Research on Terrestrial Small Mammals

Robert McCleery, Ara Monadjem, L. Mike Conner, James D. Austin, and Peter J. Taylor

 Johns Hopkins University Press

Baltimore

© 2021 Johns Hopkins University Press
All rights reserved. Published 2021
Printed in the United States of America on acid-free paper
9 8 7 6 5 4 3 2 1

Johns Hopkins University Press
2715 North Charles Street
Baltimore, Maryland 21218-4363
www.press.jhu.edu

Library of Congress Cataloging-in-Publication Data

Names: McCleery, Robert A., 1972– author.
Title: Methods for ecological research on terrestrial small mammals /
    Robert McCleery, Ara Monadjem, L. Mike Conner,
    James D. Austin, and Peter J. Taylor.
Description: Baltimore : Johns Hopkins University Press, 2021. | Includes
    bibliographical references and index.
Identifiers: LCCN 2020057149 | ISBN 9781421442112 (hardcover) | ISBN
    9781421442129 (ebook)
Subjects: LCSH: Mammals—Ecology. | Mammals—Research—Methodology.
Classification: LCC QL739.8 .M33 2021 | DDC 599—dc23
LC record available at https://lccn.loc.gov/2020057149

A catalog record for this book is available from the British Library.

*Special discounts are available for bulk purchases of this book. For more information, please
contact Special Sales at specialsales@jh.edu.*

To our wives

Contents

Acknowledgments

This book was motivated by the beauty of small mammals and the complexity of their ecological interactions, which we have toiled to understand and explicate. They have given us an unprecedented window into the natural world, challenged our thinking, aided our careers, and filled our working hours with joy; for this we are endlessly grateful. We thank our mentors—R. Honeycutt, N. Silvy, B. Leopold, M. R. Perrin, G. Bronner, J. U. M. Jarvis, I. L. Rautenbach, and the late J. A. J. Meester—for helping to spark our passion for small mammals and for teaching us many of the ideas found in this book. In turn, we are inspired by the hard work, creativity, and enthusiasm our students and technicians have demonstrated while working with small mammals. We thank our collaborators for sharing their knowledge and ideas and for making field work so enjoyable. We have also received considerable support from our institutions and supervisors, who have giving us the freedom to pursue our research interests. Finally, we thank our wives and families for tolerating dead animals in their freezers, misplaced bait bags in the laundry, and our endless weeks in the field.

Methods for Ecological Research on Terrestrial Small Mammals

1. Introduction

The term *small mammal* as generally used describes roughly three-quarters of the species in the class Mammalia. The orders Rodentia (rodents) and Chiroptera (bats), which are composed almost entirely of small mammals, alone account for >60% of extant mammals. However, because the methodologies and approaches used to study bats and nonvolant mammals are quite different, and because there are authoritative works detailing the study of bats (e.g., Kunz and Parson 2009), our focus with this text is to detail ecological methods for the study of terrestrial small mammals. Accordingly, we use *small mammal* to mean nonvolant mammals with body mass <3 kg. More specifically, the methodologies presented in this text are highly applicable to the study of species from the orders Rodentia, Afrosoricida (tenrecs, golden moles, otter-shrews), Eulipotyphla (shrews and moles), Lagomorpha (rabbits, hares, and pikas), and Didelphimorphia (opossums), but we specifically exclude the Carnivora and Primates. While there are many primates and carnivores that are <3 kg, they are ecologically distinct from other terrestrial small mammals, and many of the methods used to study smaller primates and carnivores are specialized as well as similar to the approaches used on larger members of their family.

Unlike their larger relatives, small mammals do not often capture the public's imagination, despite their equally critical role in the proper ecological functioning of most terrestrial systems. Ecologically, small mammals are important as herbivores, seed predators and dispersers, pollinators, nutrient cyclers, soil aerators, ecosystem engineers, and prey for numerous other species (Lacher et al. 2016). Small mammals also are significant in the well-being and livelihoods of human populations. Small mammals are hosts to pathogens that can be transferred to humans and cause disease. These diseases have reshaped the course of human history (e.g., plague) and to this day endanger our public health. Some small mammals are also notorious crop pests that can damage most types of crops (e.g., cereals, fruits, vegetables), causing billions of dollars of damage and loss for millions of people across the world.

Historically, there have been several resources to help guide researchers and managers interested in conducting ecological research on small mammals. Some are broad texts that cover study design, field methods, and analysis of wildlife research. These venerable texts include Charlie Krebs's *Ecological Methodology* (2014), various editions of *The Wildlife Techniques Manual* (Braun 2005, Silvy 2020), and Bill Sutherland's *Ecological Census Techniques* (2006). Such general texts are invaluable resources that we refer to regularly in this book, but their breadth precludes detailed description of many of the methodologies that are specific to the study of small mammals. There are also several texts—for example, *Field Methods for Rodent Studies in Asia and the Indo-Pacific*, by Ken Aplin and his colleagues (2003), and *Manual for Analysis of Rodent Populations*, written by David Davis in 1956—that covered some of the traditional methods used to conduct ecological field research on small mammals. These important works are not, however, up-to-date with the latest passive, molecular, and statistical approaches and can be limited to specific taxa (e.g., rodents) and regions of the world. Still, there is a treasure trove of manuscripts and book chapters detailing innovative ways to study small mammals that have yet to be compiled and synthesized.

Accordingly, there was a real need for a comprehensive, up-to-date guide on how to study the ecology of small mammals that covered the latest techniques, as well as the tried and tested methods for conducting rigorous scientific inquiry on small mammals. We crafted this book to fill that void and provide researchers, students, and managers with the information that they will need to conduct innovative, state-of-the-art research on small mammals. We highlight approaches that are currently being used for small mammal ecological research while introducing emerging technologies that might become commonplace in the coming decade. We also made a concerted effort to include research and examples from different parts of the world. We were keenly aware that the realities, challenges, and methodologies used to research small mammals in, say, deserts are quite different from those facing researchers in tropical forests. In an effort to capture the diversity of approaches used to conduct ecological research on small mammals, we incorporate research methods from different continents and ecoregions (e.g., rainforests, temperate forests, savannas, wetlands, and arid systems).

In addition to providing for the scientific community, we were compelled to write this book for selfish reasons. We wanted to have a guide that we could use, along with our students, as we developed new ways of understanding small mammals' interactions with their environments. We have always been excited about the endless opportunities that small mammals present as models for the study of the natural world. They have several traits that make them ideal for ecological research. Small mammals can be relatively easy to capture and detect, and because they have short lifespans and high fecundity rates, their populations respond rapidly to environmental changes. Additionally, because many small mammals do not move large distances and generally perceive and respond to the environment on relatively small scales, it is possible to study how everything from microhabitats to global changes alter their populations, communities, and evolutionary trajectories. Small mammals also often occur in large and complex communities, which provide opportunities to study species interaction and functional responses to change. Furthermore, there is a considerable amount of work to be done, relative to the number of species; small mammals are vastly understudied relative to their larger counterparts. In fact, for groups like rodents, we are regularly discovering new species and reconfiguring our understanding of their origins and taxonomic relationships. Lastly, in our completely biased opinion and contrary to popular belief, small mammals are cute, adorable, and fascinating animals that bring joy to those who work on them.

For those of you who may be new to small mammal research, you will not only find exciting avenues for research and ample opportunities—you are also likely to find a welcoming and supportive community. In our combined experience, we have found the community of small mammal researchers, practitioners, and enthusiasts to be welcoming and eager to share information and ideas. While there are scientific disagreements, they are rarely acrimonious. Instead, we have found that the community is eager to support any study that advances our understanding and appreciation of small mammals.

Before jumping into small mammal research, there are several important challenges that should be noted. For example, when working in areas and on species with potentially harmful pathogens, researchers should take extra precaution when directly handling or sampling small mammals. The diminutive size of small mammals also necessitates caution, precision, and practice when marking, drawing blood, or otherwise nondestructively sampling. Furthermore, while some small mammals can be easy to capture, others can be notoriously elusive and trap shy, requiring patience and creativity to study them. In some areas of the world, you are also likely to struggle even to just identify what species you have captured, either because there are no current field guides or because there is simply not enough information to make an identification. It can also be difficult to observe small mammals in the wild for behavioral research. With a few exceptions (e.g., squirrels and lagomorphs), small mammals are predominantly nocturnal, secretive, or both, and a decent

Fig. 2.2. Tracking tubes illustrating presence of *Peromyscus polionotus* in dune habitats. (*A*) The tube is raised off the ground with a dowel to ease entry; the opposite side has a removable cap into which a strip of paper with an ink pad and bait are placed. (*B*) *P. polionotus* tracks can be seen on the paper pad. *(Photo: Jeff Gore)*

of small mammals (Wilkinson et al. 2012) and to examine the effects of roads, edges, corridors, and other environmental influences on rodent populations and communities (Lapolla and Barrett 1993, Moenting and Morris 2006, Rytwinski and Fahrig 2007).

There are clear financial advantages to using tracks to study small mammals. Nonetheless, these methods generally only yield a measure of activity or relative abundance. Additionally, regardless of the substrate used, it takes considerable expertise to identify tracks to species because the tracks are often highly variable, faint, or overlapping, making them difficult to distinguish (Russell et al. 2009). To improve track identification, Russell et al. (2009) used binarized digital images of tracks and an image recognition algorithm to successfully identify three congeneric murid rodents in New Zealand.

Droppings and Conspicuous Signs

Some small mammals leave conspicuous signs in the environment that can be used to indicate their presence. Researchers have taken advantage of this by systematically searching for signs of small mammals to examine distributions and relative abundance. Nests of rodents vary in size, shape, and placement in the environment, but some are easily detectable. For example, researchers have used systemic nest searches to study squirrels (*Sciurus* spp.) (Fitzgibbon 1993), round-tailed muskrats (*Neofiber alleni*) (Faller and McCleery 2017), harvest mice (*Micromys minutus*) (Harris and Yalden 2004), woodrats (*Neotoma* spp.) (Sakai and Noon 1993), bush rats (*Otomys unisulcatus*), whistling rats (*Parotomys* spp.) (Du Plessis et al. 1991, Jackso et al. 2002), and others. Similarly, some small mammals like prairie dogs (*Cynomys* spp.), giant kangaroo rats (*Dipodomys ingens*), and southern African ground squirrels (*Geosciurus* spp.) have such large or extensive burrow systems (or colonies) that some of them can even be surveyed from the air (Sidle et al. 2001, Bean et al. 2012). Fossorial mammals can also be identified and surveyed from the mounds of dirt they excavate from their burrow systems and foraging tunnels. For example, pocket gophers (*Geomys* spp.), golden moles (Chrysochloridae), and red vizcacha rats (*Tympanoctomys barrerae*) have all been detected and surveyed by their tunnels or mounds (De Graaff and Nel 1970, Omeilia et al. 1982, Ojeda et al. 1996, Jackson et al. 2008). In open environments it is even possible to survey the mounds of fossorial mammals using Google Earth (R. Gitzen, Auburn University, unpublished data). One assumption when using mounds or tunnels as proxy for density is that more mounds and tunnels indicate more animals. This may not be true, and it is possible that increased mounds and tunnels indicate increased activity of fewer individuals in habi-

the abundance of localized small mammals in a community, potentially due to a prey selection bias of owls (Avenant 2005, Terry 2010). Nonetheless, owl pellets can provide a valuable source of DNA that can be used for a host of different molecular techniques (see below and Chapter 15); however, there is the potential of cross-contamination of samples (Taberlet and Fumagalli 1996).

Hair Sampling Tubes

A hair tube, a PVC pipe lined with double-sided adhesive tape, can be an effective means of surveying small mammals (Suckling 1978). Small mammals freely pass through the tube leaving a hair sample behind on the tape. Researchers can then remove the hair and identify it under a microscope (Teerink 2003), measuring its size (Pocock and Jennings 2006) and comparing it to a reference collection (e.g., Keogh 1985, Mickala et al. 2020), or identify it genetically. Hair tubes can be placed in trees for arboreal small mammals (Suckling 1978) or on the ground, held in place with a stake (Scotts and Craig 1988). Hair tubes can be left in the field for extended periods of time, can collect hair from multiple species, and generally require less effort than live-trapping transects (Scotts and Craig 1988). Hair tubes have successfully been used to study numerous species of small mammal, including feathertail gliders (*Acrobates pygmaeus*), common hamsters (*Cricetus cricetus*) (Reiners et al. 2011), bush rats (*Rattus fuscipes*) (Scotts and Craig 1988), shrews (*Sorex* and *Neomys* spp.) (Pocock and Jennings 2006), and red squirrels (*Sciurus vulgaris*) (Mortelliti and Boitani 2008). Tracking tubes may be particularly good at detecting species that are difficult to live-trap (Suckling 1978, Scotts and Craig 1988, Mortelliti and Boitani 2008). However, tracking tubes often have low rates of capture and may miss species present in the community. As well, the morphological identification of hair can be a potential source of error (Lobert et al. 2001, Mills et al. 2002)—a shortcoming that can be surmounted by utilizing hairs as a source of DNA (Piggott and Taylor 2003). An additional constraint of hair tubes, especially in diverse communities, is that multiple species can use a tube, requiring researchers to examine each hair individually. Modifying hair tubes to close after one visitation could limit this concern (Chiron et al. 2018). Chiron et al. (2018) conducted a quantitative assessment of hair tubes designed to monitor common small mammals by comparing results to live-trapping data. They concluded that preliminary studies should combine hair tubes and traditional trapping in order to increase the efficiency of identification based on hairs and to determine trap bias.

Dogs

The use of dogs in wildlife research and management is not new but has received increased attention as an effective tool for detecting wildlife. Dogs can detect scents 100 million times better than humans and can survey areas up to four times faster (Syrotuck 1972, Mecozzi and Guthery 2008, Dahlgren et al. 2012). Hence, dogs have been widely used to detect carnivores and their scat but have only rarely been used in the detection of small mammals. However, when dogs have been used to find rodents, they have been highly successful. Researchers found that in less than an hour dogs could detect a rare ground squirrel (*Poliocitellus franklinii*) at the same rate as two days of live-trapping (Duggan et al. 2011). Dogs also appear to be effective at locating inconspicuous rodents in putatively pest-free areas (Gsell et al. 2010). The use of such dogs certainly has limitations, which stem from extensive training, variation in individual dog's performance, a dog's prey drive (which may be driven at least in part by hunger), and their health and upkeep (Dahlgren et al. 2012). Nonetheless, we believe this noninvasive technique holds considerable potential to be applied more broadly to small mammal research.

Acoustics

Acoustic detectors are a burgeoning technology in wildlife and ecology research. Acoustics detectors are now routinely used to study populations and behaviors of birds, bats, frogs, and marine mammals (Gilley 2013). However, they have been used only sparingly in the study of small mammals—often to determine the distribution of rarer arboreal species (Adamík et al. 2019, Diggins et al. 2020).

Small mammals are relatively quiet compared with other vertebrates, but shrews and some murid rodents have been shown to use ultrasonic vocalizations (Forsman and Malmquist 1988, Kalcounis-Rueppell et al. 2006), and some small mammals, such as whistling rats (*Parotomys* spp.), chinchilla (Chinchillidae), pika (Ochotonidae), and many species of squirrels (Sciuridae), even make loud audible calls. These vocalizations would allow for the study of communication, competition, movement, population size, and demographic rates (Kalcounis-Rueppell et al. 2010, Blumstein et al. 2011, Wood et al. 2020). Acoustic samples may even provide a noninvasive way to identify and mark individual rodents based on their unique call characteristics (McCowan and Hooper 2002, Pollard et al. 2010). Acoustic technology also has considerable potential to advance our understanding of small mammal behavior. It is believed that ultrasonic communications in rodents are used

Cameras can also malfunction, particularly in harsh environments (heat, humidity, freezing temperatures, etc.), and they can be disturbed by nontarget animals. In Africa, for example, cameras are commonly lost to elephants (*Loxodonta africana*) and hyenas (Hyaenidae). Accordingly, manufactured or custom housing for camera traps can be used to reduce these potential pitfalls (McCleery et al. 2014). Also, realize that, once you have acquired all your camera trapping data, it will require a considerable amount of time to sort and process. There are a number of software platforms to aid in the sorting, databasing, and processing of pictures. Young et al. (2018) provide an excellent review and comparison of these various platforms that would be useful to anyone starting a new camera trapping project.

Baited versus Unbaited

For most passive methods, researchers will have to decide if they want to use bait or a lure to attract animals. For animals that occur at low densities or cannot be detected without an attractant, bait is a logical way to increase detections. However, using bait may bias some kinds of data, including sex ratios, age classes, behaviors, and community composition (McCoy et al. 2011). A study using cameras to assess small mammals found that bait (peanut butter and oats) increased detection of some species but not of others, leading to potentially biased measures of community composition (Paull et al. 2011). Hence, there are at least two important questions that should be addressed prior to making this decision: (1) Does bait change a small mammal's use of the environment, and at what scale? (2) Does bait influence activity levels equally for all individuals in the population (such as different sexes or age groups) and in all environments? In many cases, these questions may be irrelevant, but when they are relevant, researchers will have to decide if increasing detections from bait is worth the costs of potentially biased data.

Passive Sampling for DNA

Many small mammal species remain data deficient, particularly in the tropics, a fact that may be associated with limitations to monitoring approaches (passive or active). For example, many small mammals are difficult or impossible to differentiate based on photos, tracks, or even hair samples, exacerbating already biased or imprecise estimates of small mammal diversity (Bickford et al. 2007, Ceballos and Ehrlich 2009). Approaches that allow for passive sampling of mammal DNA are increasingly viewed as powerful tools for monitoring and research, particularly in instances where active sampling is either ethically challenging (Powell and Proulx 2003) or limited in precision and accuracy. Many of the techniques discussed above can be adapted for DNA sampling, which can in turn be utilized for genetics studies, capture-mark-recapture, species detection or identification, and even trophic niche space. Importantly, the quality of DNA (or other molecules of potential interest) will be less than that collected from active sampling (see Chapter 3) simply due to prolonged exposure to environmental conditions (e.g., ultraviolet rays, high humidity, warmer temperatures). Despite these concerns, the application of passive sampling of DNA from small mammals should continue to increase with technological advancements and increasing familiarity.

The collection of hairs or feces for DNA is particularly valuable when, as is often the case, morphological inspection provides limited information, for example, when there is a lack of access to reference databases or limited morphological variation among species (e.g., feces from rodents of similar size). DNA can increase the information content of such samples by providing information of species, gender, and even individual identification (see Chapter 15). Passive fecal sampling can also provide better dietary and natural hormone level estimates than live capture. Many rodents will defecate as they feed, allowing for the collection of feces off the ground or even off plates;

Fig. 2.5. Paper plate (weighted down with rock) showing feces from overnight visiting rodents in Eswatini savanna

this can be an effective sampling strategy (Fig. 2.5). DNA from feces can also be used to examine diet (e.g., Khanam et al. 2016, Hawlitschek et al. 2018, Chapter 11).

Markers linked to mammal sex chromosomes allow for gender identification from noninvasive samples like feces or hair (see Chapter 15). The ability to identify individuals and their sex allows for improved accuracy for estimates of abundance and individual behavioral parameters, such as habitat and dietary preferences and movements. Molecular methods can also overcome some of the restrictions presented by more traditional passive sampling (e.g., tracks and hair morphology) by providing accurate species identification and direct individual enumeration (Taberlet et al. 1999). Most examples of these types of applications come from larger mammals (see Beja-Pereira et al. 2009), probably because of the perceived ease of live-trapping most small mammals. However, many small mammals can be difficult to capture (e.g., fossorial mammals) and to enumerate using traditional means. In some cases, these species have been well characterized using passively sampled feces and hair. Microsatellite DNA amplified from feces has been used to estimate precise population size, relatedness, and dispersal in larger mammals such as common wombats (*Vombatus ursinus*) (Banks et al. 2002a, Banks et al. 2002b), while hair was used to estimate abundance and sex of hairy-nosed wombats (*Lasiorhinus krefftii*) (Banks et al. 2003). For some species, increased accuracy of demographic estimates may be achieved from noninvasive sampling relative to live-trapping (Chapter 3). For example, Sabino-Marques et al. (2018) found that live-trapping underestimated density of Cabrera voles (*Microtus cabrerae*) relative to simultaneous noninvasive sampling.

Environmental DNA

One passive method for detecting animals that is gaining a lot of attention in animal ecology is environmental DNA (eDNA), which refers to DNA isolated from environmental samples, bypassing the need to passively sample individuals. Predominantly sampled from water for vertebrate studies (eDNA studies of microbial organisms and invertebrates have been conducted on soils and air samples), eDNA represent genomic DNA from living cells originating from organisms and extracellular DNA from decaying cells. Innovations in analyses of eDNA offer alternatives for detection and identification of rare or cryptic species in freshwater systems (Ma et al. 2016, Stewart et al. 2017). DNA that is shed or excreted into the aquatic environment from individuals during normal activity can be collected, concentrated, and polymerase chain reaction (PCR) amplified using primers targeted for specific species, or using universal primers and massively parallel sequencing (Chapter 15) to detect multiple species at one time. For mammals (and other animals), mitochondrial DNA is targeted due to its high copy number in cells (relative to nuclear DNA), and thus in the environment (Rees et al. 2014).

Relative to microbiota, or more commonly studied aquatic vertebrates like fishes and amphibians, eDNA studies focused on (nonmarine) mammals have been relatively rare. Terrestrial mammals that have been detected in eDNA studies include the masked palm civet (*Paguma larvata*), the Sunda pangolin (*Manis javanica*), and squirrels (*Callosciurus adamsi, Dremomys rufigenis*, and *Petaurista alborufus*), among other mammal species detected from water collected at salt licks (Ishige et al. 2017). Klymus et al. (2017) used metabarcoding to demonstrate that wildlife, including domestic dogs, other carnivores, and small mammals, frequently utilize uranium mine containment ponds as water source. The use and efficacy of eDNA for occupancy detection models of platypus (*Ornithorhynchus anatinus*) was explored by Lugg et al. (2018). In a direct comparison, they found that conditional probabilities of detecting platypus by eDNA from a single water sample were greater than that from a single trapping visit. In their situation, eDNA was more cost-effective than trapping.

Additional Secondary Sources of Mammal DNA

Recent additions to the researcher's toolbox include secondary sources of mammal DNA. For example, researchers have recently been able to describe mammal communities using the blood meals of mosquitos and terrestrial leeches (Schnell et al. 2012, Hoyer et al. 2017). Carrion flies (Calliphoridae and Sarcophagidae) that feed on feces, open wounds, and carcasses have also been used to describe mammal communities. Because they are not host specific, nor restricted to certain habitats, carrion flies have been used to monitor mammal biodiversity (e.g., Calvignac-Spencer et al. 2013). Recent attempts to evaluate carrion-fly-derived DNA relative to other approaches (including camera trapping, scat collection, and hair trapping) have demonstrated that the ubiquity and relative ease of sampling blowflies (Calliphoridae) makes them a highly viable option. Simultaneous sampling efforts in forest reserves have demonstrated that blowfly-derived DNA detected more (but not necessarily the same) mammal species (including small mammals) than camera traps and other methods tested (Lee et al. 2016). It also appears that the gut contents of dung beetles can be used to sample mammal communities (Gillett et al. 2016). Additionally, Taberlet and Fumagalli (1996) demonstrated that rodent and insectivore skulls collected from owl pellets could serve as a source of DNA for studying population genetics.

The main potential limitations of sampling DNA from each of these noninvasive methods are DNA quality and relative cost compared to non-DNA survey methods. Because DNA autolysis in sloughed skin, hair, or other sources is rapid when exposed to ultraviolet, higher temperatures, and increasing humidity, sample quality will vary among studies. As a result, using noninvasive DNA requires additional effort to evaluate DNA quality and to gauge the accuracy of genotyping or sequencing results. Consequences of using passively sampled DNA range from simply failed PCR, generally due to template DNA being overly fragmented, to (more importantly) misleading results due to allelic dropout or misidentification due to cross-species contamination.

Ultimately, the consequences of using poor-quality DNA to generate genetic information depend on the objective of the research or monitoring program. Projects attempting to identify individuals (e.g., mark-recapture studies, estimating relatedness) can be particularly affected. Studies looking at population genetic structure may be less severely affected, depending on the extent and severity of genotyping error. Low genotyping error (~4%) was reported using DNA from common hamster (*Cricetus cricetus*) hairs (Reiners et al. 2011). Genotyping from too few hairs can lead to high error rate; for alpine marmots (*Marmota marmota*), as many as 10 hairs were needed to reduce genotyping error to acceptable levels (Goossens et al. 1998). For studies involving hair sampling, follicle quality is an important aspect predicting PCR success.

Most of the critical procedures related to working with noninvasive DNA occur once out of the field. Common issues that need to be addressed in the genetic laboratory include the increased risk of cross-contamination from other sources of DNA, further degradation of DNA (e.g., from multiple freezings and thawings of samples), a lack of ultra-clean conditions during PCR preparation leading to contamination, and ultimately allele scoring and testing for null alleles and genotyping error rates. Beja-Pereira et al. (2009) provide an excellent overview of recommendations for each step in the process of working with noninvasive DNA, including preserving, extracting, PCR, and testing for error and null alleles. Additionally, pilot studies are critical when considering carrying out noninvasive sampling. This applies to both the field and laboratory stages, in order to determine whether the technology is appropriate and to allow a cost-benefit analysis to be carried out before initiating a large-scale project. We recommend reading Barnes and Turner (2016) for a comprehensive work on using eDNA in ecological and conservation applications.

Conclusions and Future Directions

Small mammal ecologists have traditionally used live captures to understand how communities, populations, and individuals interact with the environment and with each other. However, as detailed in this chapter, there are a host of effective and efficient means to study small mammals without having to trap them. Each method that we have discussed typically works better under certain conditions or for certain questions than others. In our experience, we have also found that a method that has worked for a researcher in one part of the world has not always worked for someone in another system. For these reasons, and regardless of the passive methodology you employ, we encourage you to test your method and troubleshoot before implementing it in the field.

The passive study of wildlife has been an exciting area of growth with rapidly developing technologies. In the coming decades, there is likely to be considerably more growth in the use of acoustic surveys and camera trapping in the study of small mammal ecology. The use of these technologies will likely be followed by the increased use and precision of algorithms to detect and identify species and individuals from pictures and sound files. Although dogs have been underutilized in the study of small mammals, they appear to have tremendous potential for the study of rare species. Finally, noninvasive genotyping is an exciting and increasingly valuable and powerful tool that will see increased application for the many small mammal species that are cryptic or endangered. However, the limitations and costs of capturing animals need to be weighed against those relating to laboratory analysis of noninvasive sources of DNA.

REFERENCES

Abramsky, Z., M. L. Rosenzweig, and B. Pinshow. 1991. "The shape of a gerbil isocline measured using principles of optimal habitat selection." *Ecology* 72 (1): 329–340. https://doi.org/10.2307/1938926.

Adamík, P., L. Poledník, K. Poledníková, and D. Romportl. 2019. "Mapping an elusive arboreal rodent: Combining nocturnal acoustic surveys and citizen science data extends the known distribution of the edible dormouse (*Glis glis*) in the Czech Republic." *Mammalian Biology* 99 (1): 12–18.

Avenant, N. L. 2005. "Barn owl pellets: A useful tool for monitoring small mammal communities." *Belgian Journal of Zoology* 135 (suppl): 39–43.

Banks, S. C., S. D. Hoyle, A. B. Horsup, P. Sunnucks, and A. C. Taylor. 2003. "Demographic monitoring of an entire species by genetic analysis of non-invasively collected material." *Animal Conservation* 6: 101–107.

Banks, S. C., M. P. Piggott, B. D. Hansen, N. A. Robinson, and A. C. Taylor. 2002a.

"Wombat coprogenetics: Enumerating a common wombat population by microsatellite analysis of faecal DNA." *Australian Journal of Zoology* 50 (2): 193–204.

Banks, S. C., L. F. Skerratt, and A. C. Taylor. 2002b. "Female dispersal and relatedness structure in common wombats (*Vombatus ursinus*)." *Journal of Zoology* 256 (3): 389–399. https://doi.org/10.1017/S0952836902000432.

Barnes, M. A., and C. R. Turner. 2016. "The ecology of environmental DNA and implications for conservation genetics." *Conservation Genetics* 17: 1–17.

Bates, G. H. 1950. "Track making by man and domestic animals." *Journal of Animal Ecology* 19 (1): 21–28. https://doi.org/10.2307/1568.

Bean, W. T., R. Stafford, L. R. Prugh, H. S. Butterfield, and J. S. Brashares. 2012. "An evaluation of monitoring methods for the endangered giant kangaroo rat." *Wildlife Society Bulletin* 36 (3): 587–593. https://doi.org/10.1002/wsb.171.

Beja-Pereira, A., R. Oliveira, P. C. Alves, M. K. Schwartz, and G. Luikart. 2009. "Advancing ecological understandings through technological transformations in noninvasive genetics." *Molecular Ecology Resources* 9 (5): 1279–1301.

Bickford, D., D. J. Lohman, N. S. Sodhi, P. K. L. Ng, R. Meier, K. Winker, K. K. Ingram, and I. Das. 2007. "Cryptic species as a window on diversity and conservation." *Trends in Ecology & Evolution* 22 (3): 148–155.

Blumstein, D. T., D. J. Mennill, P. Clemins, L. Girod, K. Yao, G. Patricelli, et al. 2011. "Acoustic monitoring in terrestrial environments using microphone arrays: Applications, technological considerations and prospectus." *Journal of Applied Ecology* 48 (3): 758–767.

Bond, A. R., and D. N. Jones. 2008. "Temporal trends in use of fauna-friendly underpasses and overpasses." *Wildlife Research* 35 (2): 103–112. https://doi.org/10.1071/wr07027.

Branchi, I., D. Santucci, and E. Alleva. 2001. "Ultrasonic vocalisation emitted by infant rodents: A tool for assessment of neurobehavioural development." *Behavioural Brain Research* 125 (1–2): 49–56. https://doi.org/10.1016/s0166-4328(01)00277-7.

Calvignac-Spencer, S., K. Merkel, N. Kutzner, H. Kühl, C. Boesch, P. M. Kappeler, S. Metzger, G. Schubert, and F. H. Leendertz. 2013. "Carrion fly-derived DNA as a tool for comprehensive and cost-effective assessment of mammalian biodiversity." *Molecular Ecology* 22 (4): 915–924.

Caughley, J., C. Donkin, and K. Strong. 1998. "Managing mouse plagues in rural Australia." In *Proceedings of the Eighteenth Vertebrate Pest Conference*, edited by R. O. Baker and A. C. Crabb, 160–165. Davis: University of California.

Ceballos, G., and P. R. Ehrlich. 2009. "Discoveries of new mammal species and their implications for conservation and ecosystem services." *Proceedings of the National Academy of Sciences* 106 (10): 3841–3846.

Chiron, F., S. Hein, R. Chargé, R. Julliard, L. Martin, A. Roguet, and J. Jacob. 2018. "Validation of hair tubes for small mammal population studies." *Journal of Mammalogy* 99 (2): 478–485.

Crooks, K. R., M. Grigione, A. Scoville, and G. Scoville. 2008. "Exploratory use of track and camera surveys of mammalian carnivores in the Peloncillo and Chiricahua Mountains of southeastern Arizona." *Southwestern Naturalist* 53 (4): 510–517. https://doi.org/10.1894/cj-146.1.

Dahlgren, D. K., R. D. Elmore, D. A. Smith, A. Hurt, E. B. Arnett, and J. W. Connelly. 2012. "Use of dogs in wildlife research and management." In *The Wildlife Tech-*

niques Manual, vol. 1, 7th ed., edited by N. J. Silvy, 140–153. Baltimore: Johns Hopkins University Press.

De Bondi, N., J. G. White, M. Stevens, and R. Cooke. 2010. "A comparison of the effectiveness of camera trapping and live trapping for sampling terrestrial small-mammal communities." *Wildlife Research* 37 (6): 456–465. https://doi.org/10.1071/wrl0046.

De Graaff, G., and J. A. J. Nel. 1970. "Notes on the small mammals of the Eastern Cape National Parks." *Koedoe* 13 (1): 147–150.

Diggins, C. A., L. M. Gilley, C. A. Kelly, and W. M. Ford. 2020. "Using ultrasonic acoustics to detect cryptic flying squirrels: Effects of season and habitat quality." *Wildlife Society Bulletin*. 44 (2): 30–308. https://doi.org/10.1002/wsb.1083.

Driessen, M. M., and P. J. Jarman. 2014. "Comparison of camera trapping and live trapping of mammals in Tasmanian coastal woodland and heathland." In *Camera Trapping: Wildlife Management and Research*, edited by P. Meek, P. Fleming, G. Ballard, P. Banks, A. Claridge, J. Sanderson, and D. Swann, 253–262. Collingwood, Australia: CSIRO Publishing.

Duggan, J. M., E. J. Heske, R. L. Schooley, A. Hurt, and A. Whitelaw. 2011. "Comparing detection dog and livetrapping surveys for a cryptic rodent." *Journal of Wildlife Management* 75 (5): 1209–1217. https://doi.org/10.1002/jwmg.150.

Du Plessis, A., G. I. H. Kerley, and P. E. D. Winter. 1991. "Dietary patterns of two herbivorous rodents: *Otomys unisulcatus* and *Parotomys brantsii* in the Karoo." *South African Journal of Zoology* 26 (2): 51–54.

Engeman, R., and D. Whisson. 2006. "Using a general indexing paradigm to monitor rodent populations." *International Biodeterioration & Biodegradation* 58 (1): 2–8. https://doi.org/10.1016/j.ibiod.2006.03.004.

Faller, C. R., and R. A. McCleery. 2017. "Urban land cover decreases the occurrence of a wetland endemic mammal and its associated vegetation." *Urban Ecosystems* 20 (3): 573–580. https://doi.org/10.1007/s11252-016-0626-1.

Falzon, G., P. D. Meek, and K. Vernes. 2014. "Computer-assisted identification of small Australian rodents in camera trap imagery." In *Camera Trapping: Wildlife Management and Research*, edited by P. Meek, P. Fleming, G. Ballard, P. Banks, A. Claridge, J. Sanderson, and D. Swann, 299–306. Collingwood, Australia: CSIRO Publishing.

Feranec, R. S., E. A. Hadly, and A. Paytan. 2007. "Determining landscape use of Holocene mammals using strontium isotopes." *Oecologia* 153 (4): 943–950. https://doi.org/10.1007/s00442-007-0779-y.

Fitzgibbon, C. D. 1993. "The distribution of gray squirrel dreys in farm woodland—the influence of wood area, isolation and management." *Journal of Applied Ecology* 30 (4): 736–742. https://doi.org/10.2307/2404251.

Forsman, K. A., and M. G. Malmquist. 1988. "Evidence for echolocation in the common shrew, *Sorex araneus*." *Journal of Zoology* 216: 655–662.

Gillett, C. P. D. T., A. J. Johnson, I. Barr, and J. Hulcr. 2016. "Metagenomic sequencing of dung beetle intestinal contents directly detects and identifies mammalian fauna." *bioRxiv*: 074849.

Gilley, L. M. 2013. "Discovery and characterization of high-frequency calls in North American flying squirrels (*Glaucomys sabrinus* and *G. volans*): Implications for ecology, behavior, and conservation" (PhD diss., Auburn University). http://etd.auburn.edu/handle/10415/3480.

Glen, A. S., S. Cockburn, M. Nichols, J. Ekanayake, and B. Warburton. 2013. "Optimising camera traps for monitoring small mammals." *PLoS ONE* 8 (6). https://doi.org/10.1371/journal.pone.0067940.

Goertz, J. W., R. M. Dawson, and E. E. Mowbray. 1975. "Response to nest boxes and reproduction by *Glaucomys volans* in northern Louisiana." *Journal of Mammalogy* 56 (4): 933–939. https://doi.org/10.2307/1379671.

Goldingay, R. L., N. N. Rueegger, M. J. Grimson, and B. D. Taylor. 2015. "Specific nest box designs can improve habitat restoration for cavity-dependent arboreal mammals." *Restoration Ecology* 23 (4): 482–490. https://doi.org/10.1111/rec.12208.

Goossens, B., L. Graziani, L. P. Waits, E. Farand, S. Magnolon, J. Coulon, et al. 1998. "Extra-pair paternity in the monogamous Alpine marmot revealed by nuclear DNA microsatellite analysis." *Behavioral Ecology and Sociobiology* 43 (4–5): 281–288.

Greene, D. U., R. A. McCleery, L. M. Wagner, and E. P. Garrison. 2016. "A comparison of four survey methods for detecting fox squirrels in the southeastern United States." *Journal of Fish and Wildlife Management* 7 (1): 99–106. https://doi.org/10.3996/082015-jfwm-080.

Greene, E., R. G. Anthony, V. Marr, and R. Morgan. 2009. "Abundance and habitat associations of Washington ground squirrels in the Columbian Basin, Oregon." *American Midland Naturalist* 162 (1): 29–42. https://doi.org/10.1674/0003-0031-162.1.29.

Gsell, A., J. Innes, P. de Monchy, and D. Brunton. 2010. "The success of using trained dogs to locate sparse rodents in pest-free sanctuaries." *Wildlife Research* 37 (1): 39–46. https://doi.org/10.1071/wr09117.

Hadly, E. A. 1999. "Fidelity of terrestrial vertebrate fossils to a modern ecosystem." *Palaeogeography, Palaeoclimatology, Palaeoecology* 149 (1–4): 389–409. https://doi.org/10.1016/s0031-0182(98)00214-4.

Harris, S., and D. W. Yalden. 2004. "An integrated monitoring programme for terrestrial mammals in Britain." *Mammal Review* 34 (1–2): 157–167. https://doi.org/10.1046/j.0305-1838.2003.00030.x.

Hawlitschek, O., A. Fernández-González, A. Balmori-de la Puente, and J. Castresana. 2018. "A pipeline for metabarcoding and diet analysis from fecal samples developed for a small semi-aquatic mammal." *PLoS ONE* 13 (8): e0201763. https://doi.org/10.1371/journal.pone.0201763.

Hoyer, I. J, E. M. Blosser, C. Acevedo, A. C. Thompson, L. E. Reeves, and N. D. Burkett-Cadena. 2017. "Mammal decline, linked to invasive Burmese python, shifts host use of vector mosquito towards reservoir hosts of a zoonotic disease." *Biology Letters* 13 (10): 20170353.

Ishige, T., M. Miya, M. Ushio, T. Sado, M. Ushioda, K. Maebashi, et al. 2017. "Tropical-forest mammals as detected by environmental DNA at natural saltlicks in Borneo." *Biological Conservation* 210 (2017): 281–285.

Jackso, T. P., T. J. Roper, L. Conradt, M. J. Jackson, and N. C. Bennett. 2002. "Alternative refuge strategies and their relation to thermophysiology in two sympatric rodents, *Parotomys brantsii* and *Otomys unisulcatus*." *Journal of Arid Environments* 51 (1): 21–34.

Jackson, C. R., T. H. Setsaas, M. R. Robertson, and N. C. Bennett. 2008. "Ecological variables governing habitat suitability and the distribution of the endangered Juliana's golden mole." *African Zoology* 43 (2): 245–255. https://doi.org/10.3377/1562-7020-43.2.245.

Kalcounis-Rueppell, M. C., J. D. Metheny, and M. J. Vonhof. 2006. "Production of ultrasonic vocalizations by *Peromyscus* mice in the wild." *Frontiers in Zoology* 3 (1): 3.

Kalcounis-Rueppell, M. C., R. Petric, J. R. Briggs, C. Carney, M. M. Marshall, J. T. Willse, et al. 2010. "Differences in Ultrasonic Vocalizations between Wild and Laboratory California Mice (*Peromyscus californicus*)." *PLoS ONE* 5 (3). https://doi.org/10.1371/journal.pone.0009705.

Keeping, D. 2014. "Rapid assessment of wildlife abundance: Estimating animal density with track counts using body mass–day range scaling rules." *Animal Conservation* 17 (5): 486–497.

Keogh, H. 1985. "A photographic reference system based on the cuticular scale patterns and groove of the hair of 44 species of southern African Cricetidae and Muridae." *South African Journal of Wildlife Research* 15 (4): 109–159.

Khanam, S., R. Howitt, M. Mushtaq, and J. C. Russell. 2016. "Diet analysis of small mammal pests: A comparison of molecular and microhistological methods." *Integrative Zoology* 11 (2): 98–110.

King, C. M., and R. L. Edgar. 1977. "Techniques for trapping and tracking stoats (*Mustela erminea*); a review, and a new system." *New Zealand Journal of Zoology* 4 (2): 193–212.

Klymus, K. E., C. A. Richter, N. Thompson, and J. E. 2017. "Metabarcoding of environmental DNA samples to explore the use of uranium mine containment ponds as a water source for wildlife." *Diversity* 9 (4): 54. https://doi.org/10.3390/d9040054.

Kotler, B. P. 1985. "Microhabitat utilization in desert rodents: A comparison of two methods of measurement." *Journal of Mammalogy* 66 (2): 374–378. https://doi.org/10.2307/1381252.

Kotler, B. P., J. S. Brown, and A. Subach. 1993. "Mechanisms of species coexistence of optimal foragers: Temporal partitioning by two species of sand dune gerbils." *Oikos* 67 (3): 548–556. https://doi.org/10.2307/3545367.

Lambert, M., F. Bellamy, R. Budgey, R. Callaby, J. Coats, and J. Talling. 2017. "Validating activity indices from camera traps for commensal rodents and other wildlife in and around farm buildings." *Pest Management Science*. https://doi.org/10.1002/ps.4668.

Lapolla, V. N., and G. W. Barrett. 1993. "Effects of corridor width and presence on the population-dynamics of the meadow vole (*Microtus pennsylvanicus*)." *Landscape Ecology* 8 (1): 25–37.

Lazenby, B. T., N. J. Mooney, and C. R. Dickman. 2015. "Detecting species interactions using remote cameras: Effects on small mammals of predators, conspecifics, and climate." *Ecosphere* 6 (12). https://doi.org/10.1890/es14-00522.1.

Lee, P. S., H. M. Gan, G. R. Clements, and J. J. Wilson. 2016. "Field calibration of blowfly-derived DNA against traditional methods for assessing mammal diversity in tropical forests." *Genome* 59 (11): 1008–1022.

Lobert, B., L. Lumsden, H. Brunner, and B. Triggs. 2001. "An assessment of the accuracy and reliability of hair identification of south-east Australian mammals." *Wildlife Research* 28 (6): 637–641. https://doi.org/10.1071/wr00124.

Loggins, R. E., J. A. Gore, L. L. Brown, L. A. Slaby, and E. H. Leone. 2010. "A modified track tube for detecting beach mice." *Journal of Wildlife Management* 74 (5): 1154–1159. https://doi.org/10.2193/2009-294.

Lugg, W. H., J. Griffiths, A. R. van Rooyen, A. R. Weeks, and R. Tingley. 2018. "Optimal survey designs for environmental DNA sampling." *Methods in Ecology and Evolution* 9 (4): 1049–1059.

Ma, H., K. Stewart, S. Lougheed, J. Zheng, Y. Wang, and J. Zhao. 2016. "Characterization, optimization, and validation of environmental DNA (eDNA) markers to detect an endangered aquatic mammal." *Conservation genetics resources* 8 (4): 561–568.

Mabee, T. J. 1998. "A weather-resistant tracking tube for small mammals." *Wildlife Society Bulletin* 26 (3): 571–574.

Mahlaba, T. A. M., A. Monadjem, R. McCleery, and S. R. Belmain. 2017. "Domestic cats and dogs create a landscape of fear for pest rodents around rural homesteads." *PLoS ONE* 12 (2). https://doi.org/10.1371/journal.pone.0171593.

McCleery, R. A. 2009. "Reproduction, juvenile survival and retention in an urban fox squirrel population." *Urban Ecosystems* 12 (2): 177–184.

McCleery, R. A., and C. L. Zweig. 2016. "Leveraging limited information to understand ecological relationships of endangered Florida salt marsh vole." *Journal of Mammalogy* 97 (4): 1249–1255. https://doi.org/10.1093/jmammal/gyw084.

McCleery, R. A., C. L. Zweig, M. A. Desa, R. Hunt, W. M. Kitchens, and H. F. Percival. 2014. "A novel method for camera-trapping small mammals." *Wildlife Society Bulletin* 38 (4): 887–891. https://doi.org/10.1002/wsb.447.

McCowan, B., and S. L. Hooper. 2002. "Individual acoustic variation in Belding's ground squirrel alarm chirps in the High Sierra Nevada." *The Journal of the Acoustical Society of America* 111 (3): 1157–1160.

McCoy, J. C., S. S. Ditchkoff, and T. D. Steury. 2011. "Bias associated with baited camera sites for assessing population characteristics of deer." *Journal of Wildlife Management* 75 (2): 472–477. https://doi.org/10.1002/jwmg.54.

McDonald, P. J., R. Brittingham, C. Nano, and R. Paltridge. 2015. "A new population of the critically endangered central rock-rat (*Zyzomys pedunculatus*) discovered in the Northern Territory." *Australian Mammalogy* 37 (1): 97–100. https://doi.org/10.1071/am14012.

Mecozzi, G. E., and F. S. Guthery. 2008. "Behavior of walk-hunters and pointing dogs during northern bobwhite hunts." *Journal of Wildlife Management* 72 (6): 1399–1404. https://doi.org/10.2193/2007-555.

Mickala, A. G., S. Ntie, and V. Nicolas. 2020. "Distinguishing Central African rodents and shrews using their hair morphology." *African Journal of Ecology*. https://doi.org/10.1111/aje.12788.

Mills, D. J., B. Harris, A. W. Claridge, and S. C. Barry. 2002. "Efficacy of hair-sampling techniques for the detection of medium-sized terrestrial mammals. I. A comparison between hair-funnels, hair-tubes and indirect signs." *Wildlife Research* 29 (4): 379–387. https://doi.org/10.1071/wr01031.

Moenting, A. E., and D. W. Morris. 2006. "Disturbance and habitat use: Is edge more important than area?" *Oikos* 115 (1): 23–32.

Mortelliti, A., and L. Boitani. 2008. "Inferring red squirrel (*Sciurus vulgaris*) absence with hair tubes surveys: A sampling protocol." *European Journal of Wildlife Research* 54 (2): 353–356. https://doi.org/10.1007/s10344-007-0135-x.

Ojeda, R. A., J. M. Gonnet, C. E. Borghi, S. M. Giannoni, C. M. Campos, and G. B. Diaz. 1996. "Ecological observations of the red vizcacha rat, *Tympanoctomys barrerae* in desert habitats of Argentina." *Mastozoología Neotropical* 3: 183–191.

Omeilia, M. E., F. L. Knopf, and J. C. Lewis. 1982. "Some consequences of competition between prairie dogs and beef-cattle." *Journal of Range Management* 35 (5): 580–585. https://doi.org/10.2307/3898641.

Paull, D. J., A. W. Claridge, and S. C. Barry. 2011. "There's no accounting for taste: Bait attractants and infrared digital cameras for detecting small to medium ground-dwelling mammals." *Wildlife Research* 38 (3): 188–195. https://doi.org/10.1071/WR10203.

Petric, R., and M. C. Kalcounis-Rueppell. 2013. "Female and male adult brush mice (*Peromyscus boylii*) use ultrasonic vocalizations in the wild." *Behaviour* 150 (14): 1747–1766. https://doi.org/10.1163/1568539x-00003118.

Piggott, M. P. 2016. "Evaluating the effects of laboratory protocols on eDNA detection probability for an endangered freshwater fish." *Ecology and Evolution* 6 (9): 2739–2750.

Piggott, M. P., and A. C. Taylor. 2003. "Remote collection of animal DNA and its applications in conservation management and understanding the population biology of rare and cryptic species." *Wildlife Research* 30 (1): 1–13.

Pocock, M. J. O., and N. Jennings. 2006. "Use of hair tubes to survey for shrews: New methods for identification and quantification of abundance." *Mammal Review* 36 (4): 299–308.

Pollard, K. A., D. T. Blumstein, and S. C. Griffin. 2010. "Pre-screening acoustic and other natural signatures for use in noninvasive individual identification." *Journal of Applied Ecology* 47 (5): 1103–1109. https://doi.org/10.1111/j.1365-2664.2010.01851.x.

Potts, W. K., C. J. Manning, and E. K. Wakeland. 1991. "Mating patterns in seminatural populations of mice influenced by MHC genotype." *Nature* 352 (6336): 619–621. https://doi.org/10.1038/352619a0.

Powell, R. A., and G. Proulx. 2003. "Trapping and marking terrestrial mammals for research: Integrating ethics, performance criteria, techniques, and common sense." *ILAR Journal* 44 (4): 259–276.

Proulx, G., and M. W. Barrett. 1989. "Animal welfare concerns and wildlife trapping: Ethics, standards and commitments." *Transactions of the Western Section of the Wildlife Society* 25: 1–6.

Qamar, Q. Z., U. Ali, R. A. Minhas, N. I. Dar, and M. Anwar. 2012. "New distribution information on woolly flying squirrel (*Eupetaurus cinereus Thomas*, 1888) in Neelum Valley of Azad Jammu and Kashmir, Pakistan." *Pakistan Journal of Zoology* 44 (5): 1333–1342.

Rabiu, S., and R. K. Rose. 2019. "Demographic response of the Gambian gerbil to seasonal changes in Savannah fallow fields." *Folia Oecologica* 46 (1): 1–9.

Rees, H. C., B. C. Maddison, D. J. Middleditch, J. R. M. Patmore, and K. C. Gough. 2014. "The detection of aquatic animal species using environmental DNA—a review of eDNA as a survey tool in ecology." *Journal of Applied Ecology* 51 (5): 1450–1459.

Reiners, T. E., J. A. Encarnação, and V. Wolters. 2011. "An optimized hair trap for noninvasive genetic studies of small cryptic mammals." *European Journal of Wildlife Research* 57 (4): 991–995.

Rowcliffe, J. M., R. Kays, B. Kranstauber, C. Carbone, and P. A. Jansen. 2014. "Quantifying levels of animal activity using camera trap data." *Methods in Ecology and Evolution* 5 (11): 1170–1179. https://doi.org/10.1111/2041-210x.12278.

Russell, J. C., N. Hasler, R. Klette, and B. Rosenhahn. 2009. "Automatic track recognition of footprints for identifying cryptic species." *Ecology* 90 (7): 2007–2013. https://doi.org/10.1890/08-1069.1.

Rytwinski, T., and L. Fahrig. 2007. "Effect of road density on abundance of white-footed mice." *Landscape Ecology* 22 (10): 1501–1512. https://doi.org/10.1007/s10980-007-9134-2.

Sabino-Marques, H., C. Mendes Ferreira, J. Paupério, P. Costa, S. Barbosa, C. Encarnação, et al. 2018. "Combining genetic non-invasive sampling with spatially explicit capture-recapture models for density estimation of a patchily distributed small mammal." *European Journal of Wildlife Research* 64 (4): 44.

Sakai, H. F., and B. R. Noon. 1993. "Dusky-footed woodrat abundance in different-aged forests in northwestern California." *Journal of Wildlife Management* 57 (2): 373–382. https://doi.org/10.2307/3809436.

Schmidt, R. H., and J. G. Bruner. 1981. "A professional attitude toward humaneness." *Wildlife Society Bulletin* 9 (4): 289–291.

Schnell, I. B., P. F. Thomsen, N. Wilkinson, M. Rasmussen, L. R. D. Jensen, E. Willerslev, et al. 2012. "Screening mammal biodiversity using DNA from leeches." *Current Biology* 22 (8): R262–R263.

Schrader, C., A. Schielke, L. Ellerbroek, and R. Johne. 2012. "PCR inhibitors—occurrence, properties and removal." *Journal of Applied Microbiology* 113 (5): 1014–1026.

Scotts, D. J., and S. A. Craig. 1988. "Improved hair-sampling tube for the detection of rare mammals." *Australian Wildlife Research* 15 (4): 469–472.

Sheppe, W. 1965. "Characteristics and uses of *Peromyscus* tracking data." *Ecology* 46 (5): 630–634. https://doi.org/10.2307/1935002.

Sidle, J. G., D. H. Johnson, and B. R. Euliss. 2001. "Estimated areal extent of colonies of black-tailed prairie dogs in the northern Great Plains." *Journal of Mammalogy* 82 (4): 928–936. https://doi.org/10.1644/1545-1542(2001)082<0928:eaeoco>2.0.co;2.

Silveira, L., A. T. A. Jacomo, and J. A. F. Diniz. 2003. "Camera trap, line transect census and track surveys: A comparative evaluation." *Biological Conservation* 114 (3): 351–355. https://doi.org/10.1016/s0006-3207(03)00063-6.

Soanes, K., P. A. Vesk, and R. Van der Ree. 2015. "Monitoring the use of road-crossing structures by arboreal marsupials: Insights gained from motion-triggered cameras and passive integrated transponder (PIT) tags." *Wildlife Research* 42 (3): 241–256.

Sosa, S., F. S. Dobson, C. Bordier, P. Neuhaus, C. Saraux, C. Bosson, et al. 2020 "Social stress in female Columbian ground squirrels: Density-independent effects of kin contribute to variation in fecal glucocorticoid metabolites." *Behavioral Ecology and Sociobiology* 74: 1–12.

Sovie, A. R., R. A. McCleery, R. J. Fletcher, and K. M. Hart. 2016. "Invasive pythons, not anthropogenic stressors, explain the distribution of a keystone species." *Biological Invasions* 18 (11): 3309–3318.

Stewart, K., H. Ma, J. Zheng, and J. Zhao. 2017. "Using environmental DNA to assess population-wide spatiotemporal reserve use." *Conservation Biology* 31 (5): 1173–1182.

Suckling, G. C. 1978. "A hair sampling tube for the detection of small mammals in trees." *Wildlife Research* 5 (2): 249–252.

Syrotuck, W. G. 1972. *Scent and the Scenting Dog.* Rome, New York: Arner Publishing Inc.

Taberlet, P., and L. Fumagalli. 1996. "Owl pellets as a source of DNA for genetic studies of small mammals." *Molecular Ecology* 5 (2): 301–305. https://doi.org/10.1046/j.1365-294X.1996.00084.x.

Taberlet, P., L. P. Waits, and G. Luikart. 1999. "Noninvasive genetic sampling: Look before you leap." *Trends in Ecology & Evolution* 14 (8): 323–327.

Taylor, C. A., and M. G. Raphael. 1988. "Identification of mammal tracks from sooted track stations in the pacific northwest." *California Fish and Game* 74 (1): 4–15.

Teerink, B. J. 2003. *Hair of West European Mammals: Atlas and Identification Key.* Cambridge: Cambridge University Press.

Terry, R. C. 2010. "The dead do not lie: Using skeletal remains for rapid assessment of historical small-mammal community baselines." *Proceedings of the Royal Society B-Biological Sciences* 277 (1685): 1193–1201. https://doi.org/10.1098/rspb.2009.1984.

Thompson, I. D., I. J. Davidson, S. O'Donnell, and F. Brazeau. 1989. "Use of track transects to measure the relative occurrence of some boreal mammals in uncut forest and regeneration stands." *Canadian Journal of Zoology* 67 (7): 1816–1823.

Tye, C.A., D. U. Greene, W. M. Giuliano, and R. A. McCleery. 2015. "Using camera-trap photographs to identify individual fox squirrels (*Sciurus niger*) in the Southeastern United States." *Wildlife Society Bulletin* 39: 645–650.

Villette, P., C. J. Krebs, T. S. Jung, and R. Boonstra. 2016. "Can camera trapping provide accurate estimates of small mammal (*Myodes rutilus* and *Peromyscus maniculatus*) density in the boreal forest?" *Journal of Mammalogy* 97 (1): 32–40. https://doi.org/10.1093/jmammal/gyv150.

Whisson, D. A., R. M. Engeman, and K. Collins. 2005. "Developing relative abundance techniques (RATs) for monitoring rodent populations." *Wildlife Research* 32 (3): 239–244. https://doi.org/10.1071/wr03128.

White, J. D. M., G. N. Bronner, and J. J. Midgley. 2017. "Camera-trapping and seed-labelling reveals widespread granivory and scatter-hoarding of nuts by rodents in the Fynbos Biome." *African Zoology* 52 (1): 31–41. https://doi.org/10.1080/15627020.2017.1292861.

Wilkinson, E. B., L. C. Branch, D. L. Miller, and J. A. Gore. 2012. "Use of track tubes to detect changes in abundance of beach mice." *Journal of Mammalogy* 93 (3): 791–798. https://doi.org/10.1644/11-mamm-a-251.5.

Wood, C. M., H. Klinck, M. Gustafson, J. J. Keane, S. C. Sawyer, R. J. Gutiérrez, and M. Z. Peery. 2020 "Using the ecological significance of animal vocalizations to improve inference in acoustic monitoring programs." *Conservation Biology.* https://doi.org/10.1111/cobi.13516.

Young, S., J. Rode-Margono, and R. Amin. 2018. "Software to facilitate and streamline camera trap data management: A review." *Ecology and Evolution* 8 (19): 9947–9957.

Zoeller, K. C., S. L. Steenhuisen, S. D. Johnson, and J. J. Midgley. 2016. "New evidence for mammal pollination of Protea species (Proteaceae) based on remote-camera analysis." *Australian Journal of Botany* 64 (1): 1–7. https://doi.org/10.1071/bt15111.

3. Active Detection Techniques

Background

Many ecological studies of small mammals require the capture of the target species, which in turn requires knowledge of the techniques available for effectively trapping the animal (Krebs 2014). Not surprisingly, because of the morphological and ecological diversity of small mammals (Vaughan et al. 2011), researchers use a variety of trapping techniques, with varying degrees of efficacy for different species (Smith 1968, O'Farrell et al. 1994, Gurnell and Flowerdew 2006, Nicolas and Colyn 2006, Umetsu et al. 2006). A basic distinction can be made between devices that capture animals either with no intent of harm (live trap) or with humane euthanasia (kill trap). Unfortunately, animals occasionally die in a live trap, and they may not die instantly in a kill trap, but the intention of the researcher is clearly different in deploying one or the other type of trap.

Aims

In this chapter we introduce various types of traps used in the study of small mammals. For each type of small mammal trap, we describe how the trap works. We conclude with suggestions regarding choice of traps to meet study objectives, baiting, sampling duration, and factors affecting efficacy of trapping.

Live Traps

In many ecological studies, it is undesirable to kill the animals. This is true, for example, in population studies, for species threatened with extinction, or even for the collection of most genetic materials (Chapter 4). In fact, most modern ecological research requires live capture of the focal animal. A large number of options are available for capturing small mammals alive and unharmed.

Sherman Traps

Perhaps the most widely used and popular live trap for capturing rats and mice (superfamily Muroidea) is the walk-in live trap, of which the Sherman trap (Fig. 3.1) is the most well-known. Importantly, folding Sherman traps have the distinct advantage of allowing researchers to transport and store a large number of traps in a relatively small area. Sherman live traps, used since the 1920s, are light aluminum (or occasionally steel) traps that open up into a box with spring-loaded doors at either end. One door remains shut and the opposite door is kept open by latching it to the treadle, which is attached to a trigger plate. A small mammal entering a Sherman trap will step onto the trigger plate, which moves the treadle, releasing and closing the door behind the animal.

Although the Sherman trap may be the most familiar, there are numerous variants of it available on the market, including the Elliot, Havahart, Longworth, and Ugglan traps, to mention just four alternatives. Differences in capture success have been reported for different types of traps (Table 3.1) (Jacob et al. 2002, Anthony et al. 2005, Jung 2016), so it is important to be familiar with the species of interest and the habitat or geographical location of your study since both may affect your choice of traps (see "Choice of trap" section, below).

A number of multiple-capture live trap designs are also available, for example the Ratinator Multiple Catch live trap (Rugged Ranch, Vista, California); these traps are useful when wishing to increase the capture rate. A major advantage of such a system is that the trap continues to work even after the capture of an animal, unlike traditional traps, which can only catch one animal at a time. Such designs are often used in pest control programs.

Box Traps

One type of live trap that is worth mentioning separately are "box" traps (e.g., Tomahawk, Baumgartner, and Havahart traps) (Fig. 3.2), which function in a similar way to Sherman traps but are typically larger and have sides made of wire-mesh, wood, or plastic, as opposed to sheets of metal. This makes them ideal for catching larger small mammals such as giant-sized rats (e.g., *Cricetomys* and *Phloeomys* spp.), squirrels, semi-fossorial species (e.g., *Octodon degus*), Australian marsupials, or even some species of rats and mice (Laurance 1992, Burger et al. 2009, Strauss et al. 2008). These traps are also regularly used by carnivore biologists to trap various species of smaller mammalian carnivores such as weasels and stoats (Mustelidae), genets (*Genetta* spp.), and

Fig. 3.1. Sherman live traps placed on the ground (*A*) and off the ground on lianas (*B*). Note that the Sherman trap on the ground has been placed on level terrain. Traps placed on trees or lianas need to be tied down to prevent them from falling to the ground.

Table 3.1. Comparison of some of the more commonly used or better-known trapping techniques for catching terrestrial small mammals under different conditions

Trap type	Description	Use	Target small mammals	Environments	Comments
Sherman	Foldable metal box with spring-loaded door	Live capture	Murids and other small-sized rodents	Can be used in most vegetation types but not ideal in arid (open) situations where the metal trap sits directly under a hot sun	Comes in different sizes, allowing the capture of a wide range of small mammals. Widely used by mammalogists.
Longworth	A metal trap consisting of a tunnel leading into a nest box	Live capture	Small-sized rodents	Can be used in most vegetation types but not ideal in arid (open) situations where the metal trap sits directly under a hot sun	The "tunnel" can be stored inside the "nest box," greatly reducing the trap's size for transportation. Popular in the United Kingdom.
Elliot	Foldable metal box with spring-loaded door	Live capture	Small-sized rodents, small marsupials	Can be used in most vegetation types but not ideal in arid (open) situations where the metal trap sits directly under a hot sun	The relatively large size of this trap allows the capture of a wide range of small mammals including small marsupials. Popular in Australia.
Ugglan	Metal wire-mesh box with a solid plastic floor that can be removed; a sliding metal cover protects the animal inside from the elements	Live capture	Small-sized rodents	The cover is not complete and, hence, the trapped animal may suffer from cold or wet weather conditions; not ideal in wet or cold conditions	Allows for multiple captures (at different times), but the sensitivity of the treadle cannot be adjusted. Has a shrew hole to allow shrews to escape, but may trap larger mice, which can get stuck in the hole. Used in Sweden.
Havahart	Non-folding metal trap with wire-mesh sides and top and one or two trapping doors	Live capture	Murids and other small-sized rodents, squirrels, and chipmunks	The trapped animal is more or less completely exposed to the elements; hence, this trap should probably not be used in extreme climates, especially in cold and wet conditions	Comes in a range of sizes, so it can be used to capture a broad spectrum of small mammals. Relatively bulky to transport.

cage that can be closed with the animal in it. For example, a network of drift fences together with Tomahawk traps has been used effectively to capture marsh rabbits (*Sylvilagus floridanus*) (Faulhaber et al. 2005) and to improve the efficacy of pitfall traps (see "Pitfall Traps," above).

Trap Barrier Systems

This is a relatively recent development that has proven to be more effective at controlling pest rodents in agricultural landscapes, particularly in rice fields, than traditional trapping systems (Aplin et al. 2003). Essentially, this system combines drift fences with multiple-capture traps to great effect (Fig. 3.5). The original design involves a drift fence of variable length, but often well over 100 m in length, that has openings every 10 to 20 m with a multiple-capture trap set at each opening, and alternating on different sides of the fence. In particular, this system is great for capturing trap-shy species such as *Rattus argentiventer*, which is a major pest of rice fields but is rarely captured in live traps. A major advantage of this system is that it does not rely on bait to lure the small mammals into the trap; instead, the rodents are "forced" into it by the drift fence. A modified version involves planting a small field covering an area of around 10 m × 10 m three weeks earlier than the main crop. The trap barrier system is then set up around this inner field, which acts as the lure, and hence the multiple-capture traps are all set on the inside to capture the rats as they try to gain access to the rice crop.

Others

Artificial nest boxes (Chapter 2) have been used successfully to study the population dynamics of a number of small mammals, including four didelphid marsupials in the Atlantic forests of Brazil. All four species were arboreal and rarely, if ever, captured with conventional trapping methods. However, the placement of artificial nest boxes allowed researchers to study the population parameters of all four species (Loretto and Vieira 2011).

Additionally, another set of unique small mammal traps has been designed for fossorial mammals, those that burrow underground. Most fossorial mammals are captured by breaching (e.g., causing in changes in air pressure, moisture, and temperature) their burrow system and placing a burrow trap in the exposed tunnel. When attempting to plug the hole in its burrow system, the animal enters the trap, usually cylindrical in shape, which shuts behind it. Several variants of this burrow trap exist (Baker and Williams 1972, Hart 1973, Connior and Risch 2019). However, the efficacy of all of these traps is limited because doors that shut passively can be propped open with dirt, while spring-loaded doors can potentially injure the animal as it enters (Moore et al. 2019). One creative way around this is to remove the need for a

Fig. 3.5. Trap barrier system showing the combination of drift fence (*A*) and multiple-capture traps (*B*) typically set in an agricultural setting such as a rice field. Note the rice within the trap barrier was planted several weeks earlier to lure pest species to this area. *(Photos: Grant R. Singleton)*

door by using PVC piping to funnel the small mammal into a capture bucket situated below (Fig. 3.6) (Moore et al 2019).

The live capture of fossorial small mammals in a bucket is just one example of the many ingenious ways ecologists have come up with to capture small mammals. Professor Herwig Leirs has a large collection of small mam-

Fig. 3.8. Victor mouse (snap) traps—note that neither of the two traps is set in this image

"mouse trap" and generally break the neck of the captured animal as opposed to crushing the skull. Museum special traps generally are more costly than standard snap traps.

More recently, a number of new designs of snap traps have appeared, such as the Romax Snap-R rat trap (Barrettine Environmental Health, Bristol, UK). These traps are typically made of plastic and have a more sensitive mechanism allowing for a higher capture rate.

"Gopher Traps"

Fossorial small mammals, such as gophers (Geomyidae), mole-rats (Bathyergidae), moles (Talpidae), and golden moles (Chrysochloridae), can be exceedingly difficult to live capture, as they rarely leave their burrows. Mole-rats can be dug out of their burrows, but this highly destructive means of capture is not recommended because it completely destroys the entire burrow system. In addition to the live traps presented above, there are specially designed "gopher traps" (Fig. 3.9) that can be used for the capture of many of these fossorial species, although nearly all of these traps involve the killing of the captured animal. The most basic gopher trap is the Macabee trap, which is a twisted piece of wire attached to a spring at one end and two sharpened prongs at the other. The prongs are twisted apart and the trap is then set inside the burrow of a small mammal. When the gopher, mole-rat, or mole pushes forward and over the trap, the prongs are released and they squeeze the ani-

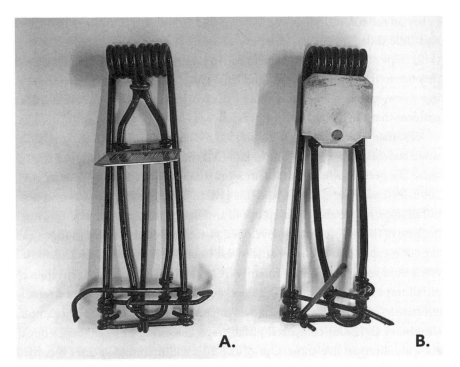

Fig. 3.9. Macabee gopher traps: (*A*) set trap; (*B*) released trap. The two prongs on either side are released when the back plate is pushed by the gopher (or other fossorial mammal) and will usually pierce the body of the animal, killing it.

A. **B.**

mal to death. There are a number of variants of this basic design, but all work in more or less the same way.

Choice of Trap

When selecting the appropriate way to capture a small mammal for a particular project, researchers have a broad array of traps and techniques to choose from. However, there are only a few options available for the trapping of most groups of small mammals. Broadly speaking, if you want to study small to medium-sized rodents (such as rats and mice), then Sherman live traps (or an equivalent) are probably appropriate. Larger rodents, such as the giant rats of Africa (*Cricetomys* spp.) and squirrels (Sciuridae) can be captured in Tomahawk traps, shrews in pitfall traps, and fossorial species in specially design burrow traps or "gopher traps." However, it is worth you spending the time to become familiar with your study animal and location to ensure that whatever method you choose is effective at capturing your study animal, is cost-effective to deploy, and causes the least amount of stress to the animal.

Be aware that trap efficacy has been tested in some parts of the world, and this may help inform your selection of trap type (Table 3.1). For example, Woodman et al. (1996) demonstrated that Victor mouse (kill) traps captured 3.5 times more individual small mammals than Sherman live traps in a

attached to the branch. Bungie cords or strips of car tire inner tube are useful and cheap in this respect.

Be aware of that the height of the trap above the ground does affect capture success (Malcolm 1991, Risch and Brady 1996, Astúa et al. 2006), and an attempt should be made to get traps up into the canopy if effective sampling of that stratum is desirable. For example, in an Amazonian study, certain arboreal species were regularly captured at 15 m above the ground, but not at 2 m above the ground, and never at ground level (Malcolm 1991). Placing traps at height can be challenging in a tropical rainforest. In the Malcom (1991) study, traps were placed high up in trees using one of two successful methods. The first involved creating a V-shaped platform by nailing wooden poles (0.9 m in length and 3 cm in diameter) to the trunk and side branch onto which the trap was fixed. The second involved attaching a pulley system to a high branch from which a platform (with attached trap) on a rope could be hoisted up. Interestingly, in forests with a discontinuous canopy, trapping within the canopy does not add to species richness and the effort of placing traps that high may not be justified (Lambert et al. 2005).

Wetlands and areas that are coastally flooded are often home to semi-aquatic small mammals such as rice rats (*Oryzomys* spp.) and water voles (*Arvicola* spp.). These areas can create challenges for researchers because changing water levels may drown animals in traps, and it can be difficult to find dry ground to place a trap. To get around this, you can secure your traps with rubber bands or Velcro to a rectangular piece of buoyant foam or insulation board. On one side of the foam, make a small round hole that you can place a long, narrow stake or dowel through. The dowel is place into the ground and the hole in the foam slides over the dowel; the trap will stay dry and move freely as the water level changes (Fig. 3.10).

Beyond these basic considerations, the placement of your traps will determine what sorts of questions that you can answer. For example, Sherman traps set in straight line transects are more efficient at capturing diversity; however, transects are of little use in determining population sizes (Pearson and Ruggiero 2003). Conversely, traps laid out in a grid fashion or in trapping webs are good for estimating population sizes but inefficient for capturing diversity. This topic is developed further in Chapter 7.

Bait and Trap Disturbance

The bait that you use determines the species that visit your trap. Hence, consideration of bait type is critical to any small mammal trapping operation. Perhaps the most familiar bait for small mammals is peanut butter and rolled oats mixed together, with perhaps some raisins added to the paste. This bait

Fig. 3.10. Sherman live trap attached to foam board, allowing it to move freely with changing water level and to keep animals dry in the trap. *(Photo: Wesley W. Boone)*

works well for rats and mice throughout much of the world, but unfortunately it is not a universal silver bullet. For example, this traditional bait does not work very well in forests of tropical Africa, where palm nut is more effective, or in Neotropical rainforests, where researchers add meat or fatty substances (for carnivorous small mammals) and fruit (for frugivores small mammals) (Malcolm 1991, Bovendorp et al. 2017).

The effectiveness of various bait types has been tested in many different environments and regions, including North American forests (Patric 1970), southern African grasslands (Kok et al. 2013), and Neotropical forests (Woodman et al. 1996, Astúa et al. 2006), to mention just four studies. Other studies have similarly tested bait efficacy elsewhere (Gurnell 1976, Paull et al. 2011, Hice and Velazco 2013, Kumar et al. 2013). The efficacy of bait not only varies with region but also with target species (Table 3.2). Therefore, one's research question may affect the type of bait selected. For example, Kok et al. (2013) showed that peanut butter and oats captured the greatest number of individuals, whereas birdseed and banana captured the greatest diversity of small mammal species. Furthermore, some species are rarely captured with traditional baits and may require a special mix. For example, vlei rats (*Otomys* spp.) hardly ever enter Sherman traps with traditional baits, even where these animals occur at high densities, but they may be captured in relatively

warmer) conditions puts the trapped small mammal at risk of hypothermia. Placing cotton balls or something similar inside the traps provides artificial bedding for the animal to snuggle up in and will assist in keeping the animal warm. Researchers should also consider the use of insulated traps with polystyrene foam or polyethylene bags (Moro 1991). Regular checks of traps are also required.

Conclusion and Future Directions

This chapter has summarized the various ways in which small mammals may be captured. Over the past few decades, hundreds of capture techniques have been developed or discovered, but most ecologists generally need only employ one or a few of these in any particular study. The goal of this chapter was to provide a range of possible options for deciding which technique will work for any particular project. While many of the techniques presented here have been highly effective in certain contexts and circumstances, it does not mean they will be effective under other circumstances. In some instances, you might need to develop your own capture method. Furthermore, as capture methods and technologies are continually being developed, it is increasingly likely that small mammals that are difficult to capture with today's technology may become easier to capture with a new or modified method.

REFERENCES

Anthony, N. M., C. A. Ribic, R. Bautz, and T. Garland. 2005. "Comparative effectiveness of Longworth and Sherman live traps." *Wildlife Society Bulletin* 33 (3): 1018–1026. https://doi.org/10.2193/0091-7648(2005)33[1018:CEOLAS]2.0.CO;2.

Aplin, K. P., P. R. Brown, J. Jacob, C. J. Krebs, and G. R. Singleton. 2003. *Field Methods for Rodent Studies in Asia and the Indo-Pacific.* ACIAR Monograph No. 100. Canberra: Australian Centre for International Agricultural Research.

Astúa, D., R. T. Moura, C. E. V. Grelle, and M. T. Fonseca. 2006. "Influence of baits, trap type and position for small mammal capture in a Brazilian lowland Atlantic forest." *Boletim Do Museu Biológica Mello Leitão* 19: 31–44.

Baker, R. J., and S.L. Williams. 1972. "A live trap for pocket gophers." *Journal of Wildlife Management* 36 (4): 1320–1322.

Barros, C. S., T. Püttker, B. T. Pinotti, and R. Pardini. 2015. "Determinants of capture-recapture success: An evaluation of trapping methods to estimate population and community parameters for Atlantic forest small mammals." *Zoologia (Curitiba)* 32 (5): 334–344. https://doi.org/10.1590/S1984-46702015000500002.

Boonstra, R., and C.J. Krebs. 1976. "The effect of odour on trap response in *Microtis townsendii*." *Journal of Zoology, London* 180: 467–476.

Bovendorp, R. S., R. A. McCleery, and M. Galetti. 2017. "Optimising sampling methods for small mammal communities in Neotropical rainforests." *Mammal Review* 47 (2): 148–158. https://doi.org/10.1111/mam.12088.

Brown, J. S, B. P. Kotler, R. J. Smith, and W. O. Wirtz. 1988. "The effects of owl predation on the foraging behavior of heteromyid rodents." *Oecologia* 76: 408–415.

Burger, J. R., A. S. Chesh, R. A. Castro, L. Ortiz Tolhuysen, I. Torre, L. A. Ebensperger, et al. 2009. "The influence of trap type on evaluating population structure of the semifossorial and social rodent *Octodon degus*." *Acta Theriologica* 54 (4): 311–320. https://doi.org/10.4098/j.at.0001-7051.047.2008.

Chitty, D., and D. A. Kempson. 1949. "Prebaiting small mammals and a new design of live trap." *Ecology* 30 (4): 536–542.

Conard, J. M., J. A. Baumgardt, P. S. Gipson, and D. P. Althoff. 2008. "The influence of trap density and sampling duration on the detection of small mammal species richness." *Acta Theriologica* 53: 143–156.

Connior, M. B., and T. S. Risch. 2019. "Life trap for pocket gophers." *The Southwestern Naturalist* 54 (1): 100–103.

Dalby, P. L., and D. O. Straney. 1976. "The relative effectiveness of two size of Sherman live traps." *Acta Theriologica* 23: 311–313.

Dizney, L., P. D. Jones, and L. A. Ruedas. 2008. "Efficacy of three types of live traps used for surveying small mammals in the Pacific Northwest." *Northwestern Naturalist* 89: 171–180. https://doi.org/10.1898/NWN08-18.1.

Edalgo, J. A., and J. T. Anderson. 2007. "Effects of prebaiting on small mammal trapping success in a morrow's honeysuckle-dominated area." *Journal of Wildlife Management* 71 (1): 246–250. https://doi.org/10.2193/2006-344.

Faulhaber, C. A, N. J. Silvy, R. R. Lopez, B. A. Porter, A. Philip, and M. J. Peterson. 2005. "Use of drift fences to capture lower keys marsh rabbits." *Wildlife Society Bulletin* 33 (3): 1160–1163.

Getz, L. L. 1961. "Responses of small mammals to live-traps and weather conditions." *American Midland Naturalist* 66 (1): 160–170.

Gurnell, J. 1976. "Studies on the effects of bait and sampling intensity on trapping and estimating wood mice, *Apodemus sylvaticus*." *Journal of Zoology, London* 178: 91–105. https://doi.org/10.1111/j.1469-7998.1976.tb02265.x.

Gurnell, J. 1980. "The effects of prebaiting live traps on catching woodland rodents." *Acta Theriologica* 25 (20): 255–264. https://doi.org/10.4098/AT.arch.80-20.

Gurnell, J., and J. R. Flowerdew. 2006. *Live Trapping Small Mammals: A Practical Guide.* 4th ed. London, UK: The Mammal Society.

Hart, E. B. 1973. "A simple and effective live trap for pocket gophers." *The American Midland Naturalist* 89 (1): 200–202.

Hice, C. L., and P. M. Velazco. 2013. "Relative effectiveness of several bait and trap types for assessing terrestrial small mammal communities in Neotropical rainforest." *Occasional Papers Museum of Texas Tech University* 316: 1–15.

Hurst, Z. M., R. A. McCleery, B. A. Collier, R. J. Fletcher Jr., N. J. Silvy, P. J. Taylor, and A. Monadjem. 2013. "Dynamic edge effects in small mammal communities across a conservation-agricultural interface in Swaziland." *PLoS ONE 8* (9): e74520. https://doi.org/10.1371/journal.pone.0074520.

Hurst, Z. M., R. A. McCleery, B. A. Collier, N. J. Silvy, P. J. Taylor, and A. Monadjem. 2014. "Linking changes in small mammal communities to ecosystem functions in

Strauss, M. W., M. Sher Shah, and M. Shobrak. 2008. "Rodent trapping in the Saja/Umm Ar-Rimth protected area of Saudi Arabia using two different trap types." *Zoology in the Middle East* 43 (1): 31–39. https://doi.org/10.1080/09397140.2008.10638266.

Stromgren, E. J., and T. P. Sullivan. 2014. "Influence of pitfall versus Longworth live-traps, bait addition, and drift fences on trap success and mortality of shrews." *Acta Theriologica* 59: 203–210. https://doi.org/10.1007/s13364-013-0149-6.

Taylor, P. J., S. Downs, A. Monadjem, S. J. Eiseb, L. S. Mulungu, A. W. Massawe, et al. 2012. "Experimental treatment-control studies of ecologically based rodent management in Africa: balancing conservation and pest management." *Wildlife Research* 39 (1): 51–61. https://doi.org/10.1071/WR11111.

Torre, I., D. Guixé, and F. Sort. 2010. "Comparing three live trapping methods for small mammal sampling in cultivated areas of NE Spain." *Hystrix* 21 (2): 147–155. https://doi.org/10.4404/Hystrix-21.2-4462.

Umetsu, F., L. Naxara, and R. Pardini. 2006. "Evaluating the efficiency of pitfall traps for sampling small mammals in the Neotropics." *Journal of Mammalogy* 87 (4): 757–765. https://doi.org/10.1644/05-MAMM-A-285R2.1.

Vaughan, T. A., J. M. Ryan, and N. J. Czaplewski. 2011. *Mammalogy.* 6th ed. Sudbury, MA: Jones and Bartlett Publishers.

Wijesinghe, M. R. 2010. "Efficiency of live trapping protocols to assess small mammal diversity in tropical rainforests of Sri Lanka." *Belgian Journal of Zoology* 140: 212–215.

Willan, K. 1986. "Bait selection in laminate-toothed rats and other southern African small mammals." *Acta Theriologica* 31 (26): 359–363.

Williams, D. F., and S. E. Braun. 1983. "Comparison of pitfall and conventional traps for sampling small mammal populations." *Journal of Wildlife Management* 47 (3): 841–845.

Williams, S. E., H. Marsh, and J. Winter. 2002. "Spatial scale, species diversity, and habitat structure: Small mammals in Australian tropical rain forest." *Ecology* 83 (5): 1317–1329. https://doi.org/10.2307/3071946.

Wirminghaus, J. O., and M. R. Perrin. 1993. "Seasonal changes in density, demography and body composition of small mammals in a southern temperate forest." *Journal of Zoology, London* 229: 303–318.

Woodman, N., R. M. Timm, N. A. Slade, and T.J. Doonan. 1996. "Comparison of traps and baits for censusing small mammals in Neotropical lowlands." *Journal of Mammalogy* 77 (1): 274–281. https://doi.org/10.2307/1382728.

Zwolak, R., and K. R. Foresman. 2007. "Effects of a stand-replacing fire on small-mammal communities in montane forest." *Canadian Journal of Zoology* 85: 815–822. https://doi.org/10.1139/Z07-065.

4. Lethal and Nonlethal Sampling for Genetics, Disease Studies, and Curatorial Collections

Background

Collections of specimens have historically played a central role in mammalogy (Rocha et al. 2014). Such specimens have formed the basis for our understanding of fundamental aspects of mammalogy, such as species ecology, evolution, and systematics, but also more recent developments associated with public health, understanding invasive species, and issues associated with global change (Suarez and Tsutsui 2004, Pyke and Ehrlich 2010). Broadly, specimens can include whole vouchers bound for museums and natural history collections, where they can provide a bounty of information for ongoing or future studies (Malaney and Cook 2018, Cook and Light 2019). They can also include partial specimens, such as tissue samples for biobanks or any other sample that might involve manipulating small mammals in the field for the purpose of obtaining a biosample.

Sampling of small mammals typically requires some form of manipulation, which will depend on the objectives of the study. Passive sampling can be employed to make observations about species ecology, such as identifying tracks or feces, or to study genetics (e.g., Ferreira et al. 2018; Chapter 2). However, when trapping is performed, considerably more information can be accessed. For example, from live-trapped animals, researchers can obtain hairs for stable isotope analysis and blood for immunoassays, and accurate phenotypic measurements can be taken from lightly anesthetized animals. In addition, whole specimens may be sacrificed for studies of pathogens or as voucher specimens (Cook et al. 2016). It is critical for researchers studying small mammals to have access to accurate and practical information for collecting specimens of small mammals.

the animal's breathing. In most cases, a partly opened cloth bag can serve as a protective layer between the researcher and animal. It is also important to minimize handling time, as many small mammals can become stressed with excessive handling. Handling time can increase when animals are marked and especially with the attachment of radiotelemetry (see Chapter 5). In such cases, researchers should consider using a customized handling bag (Koprowski 2002, McCleery et al. 2007) or they should anesthetize the animal (see below).

Similar caution should be applied to handling traps and transporting trapped mammals (e.g., from trap site to processing locations). Handling traps bare-handed exposes you to bodily fluids and the potential for bites when using mesh cages. Wearing thick rubber gloves for trap handing is a good strategy as the gloves can be easily disinfected between uses with bleach. To avoid respiratory infection and to avoid over-stressing the animal due to exposure to excessive heat and sun, trapped animals should not be transported inside passenger vehicles (truck beds are ideal).

Euthanasia and Voucher Preparation

Below we describe major steps in planning and implementing sampling with an emphasis on performing these steps under nonlaboratory conditions.

Choosing a Processing Site

When preparing specimens in field settings, it is important to remember that small mammals should not be processed in close proximity to human habitation or facilities that are for mixed use (e.g., field camp, cooking, or washing areas). Rooms used for animal processing should be dedicated for that purpose and careful disinfection should be practiced following dissections and sampling. If working outdoors, an awning or tarp setup is recommended if protection from the sun is needed. Animals queued for sampling should be kept in ventilated areas, out of the sun. Provision of water and food is also necessary if processing is not immediate.

Preparation of Equipment for Sampling

Prior to euthanizing or otherwise sampling small mammals (e.g., tissues, blood, ectoparasites), all required sampling equipment should be prepared and available to allow for timely processing. For animal dissection, a good stable work surface is required, preferably one that is nonporous and can be easily disinfected. In addition, a dissection surface (e.g., dissection tray) that will allow for easy disinfection can be purchased from scientific supply companies (e.g., Carolina Biological). Inexpensive dissecting surfaces can be

Fig. 4.2. Prepared bleach, water, and ethanol for cleaning instruments between tissues or animals

made out of polystyrene covered with thinner plastic, which will allow for pinning and easy clean up and disposal.

Prepare disinfectant for cleaning tools between specimens and, as necessary, between organ types to avoid cross-contamination. Household bleach is a readily available and effective disinfectant. For sampling tissues for genetic or other diagnostic applications, have bleach, water, and ethanol (rubbing alcohol, or EtOH) available: disinfect in bleach (10% or stronger), remove excess bleach by rinsing with water, and finally, wipe with EtOH (Fig. 4.2).

Sample storage should be organized to minimize effort and recording error (particularly when large numbers of samples are being prepared). Tubes for tissue storage should be labeled with the same identification number used for the whole specimen. Chemical-resistant markers should be used for labeling the outside of Eppendorf tubes, Falcon tubes, and associated tags. We recommend purchasing and testing different markers prior to leaving for the field. Solvent-resistant tags are useful for inserting with organs or tissues in larger Falcon tubes (e.g., XyliTAG Xylene and Chemical Resistant Labels from GA International) or for adhering to the outside of tubes (e.g., Tough-Spots labels).

Tissue samples preserved in EtOH or other preservatives (see below) can be stored in 1.5 or 2.0 mL Eppendorf tubes or other brands or styles that are rated for the temperature ultimately stored (e.g., −20°C). Eppendorf tubes with Safe-Lock will add piece of mind from accidental openings. Individual tubes can also be wrapped in Parafilm (Bemis) for transportation (a good idea for precious samples). Samples bound for immediate freezing in liquid nitro-

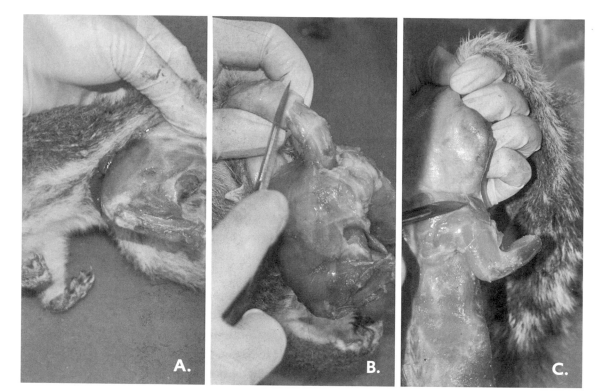

Fig. 4.5. Steps for removing intact skin from a gray squirrel: cut ventral surface and work rear legs out one at a time, removing the foot at the knee or below (*A*); peel the tail skin away from the connective tissue using a scalpel (*B*); pull the skin over shoulders (*C*)

Wire pins made from stretched Monel wire or galvanized wire of appropriate thickness are inserted into the four limbs and the tail, leaving a small section within the body. The midventral incision is then sewed up and the skin is pinned dorsally onto a cork or polystyrene board. The skull and/or postcranial skeleton are labeled separately and allowed to dry. De-fleshing of the skull and postcranial skeleton can be done using dermestid beetle colonies or by maceration, allowing the bones to soak for days or weeks (depending on their size) before removing the flesh with a fine forceps, usually under a dissecting microscope for very small specimens.

There are numerous well-illustrated sources for performing dry mounts on small mammals and for standard methods for keeping a field catalog and tagging specimens (Hall 1962, Setzer 1968, Nagorsen and Peterson 1980, Griffin and Kolberg 2004).

Liquid Preservation

In cases where there is no time to complete the skin preparations in the field, or where internal organs are required for further study, wet specimens can be retained. Wet specimens involve "pickling" the entire carcass, usually initially in 10% formalin for two weeks, followed by transferring to 70% EtOH for long-term storage (see Table 4.1). If the animal is not opened for necropsy purposes, it should be opened from the lower abdomen to the upper thorax to allow for liquid to reach all tissues (Mills, Yates, et al. 1995). Skulls can be removed from such species, either before or after pickling, by making an incision at the mouth and removing the skull. Although formalin results in complete fixation of the specimens, making them more pliable compared with pickling specimens directly in 70% EtOH (which make the specimens hard and brittle), tissues preserved in formalin are not suitable for DNA extraction. Many museums initially remove soft tissues (such as muscle tissue or liver tissue) from the fresh carcass and store this in a tube for DNA prior to pickling the carcass or making a study skin.

In general, the preservation of whole animals and tissues (not only considering genetic applications, e.g., for anatomical/histological studies) can be achieved with a range of preservative liquids (Table 4.1), such as 10% formalin or 70% EtOH. The former is a much better fixative than EtOH and results in tissues being softer and less brittle than those stored in 70% EtOH. There are human health risks associated with formaldehyde, which is a carcinogen (Morgan 1997). For organs that need to be preserved for anatomical/histological analysis, Bouin's fluid is often best. This fixative is a mixture of picric acid, 10% formalin (40% formaldehyde solution), and glacial acetic acid.

Table 4.1. Some commonly used fixatives and preservatives for small mammal tissue

Material	Ingredients	Quantity	Application
Ethyl alcohol (EtOH), 70%	95% EtOH	70 mL	Preservative, fix and harden tissue, long-term storage
	Distilled H_2O	25 mL	
Bouin's fluid	Picric acid, saturated aqueous solution	750 mL	Fixative for tissues and organs for microscopic study. For wet tissue preservation for microscopic work, fix in Bouin's solution for <24 hrs then remove excess using several 70% alcohol washes. Store wet sample in 70% alcohol.
	Formalin (40% formaldehyde solution)	250 mL	
	Glacial acetic acid	50 mL	
Embalming fluid	Formalin (40% formaldehyde solution)	5 mL	Preservative, especially for whole mammals for anatomical dissection
	Glycerin	5 mL	
	Phenol	5 mL	
	Water	85 mL	
Formalin 10% (Neutral, buffered)	Formalin (37%–40% formaldehyde stock solution)	100 mL	Preservative and fixative, whole animals, tissues and organs, indefinite storage time
	Distilled H_2O	900 mL	
	Sodium acid phosphate solution NaH_2PO_4	4 g/L	Note: 10% represents 10% of stock solution; actual amount of formaldehyde in final vol (1L) ~4%
	Anhydrous disodium phosphate Na_2HPO_4	6.5 g/L	

using a scalpel or needle, as a lancet has a higher rate of success and causes less pain and tissue trauma to the rodent (Golde et al. 2005).

Blood should be immediately collected, as blood flow is rapid and ceases after a short time. While holding the rodent by the skin between the shoulders, insert the lancet to the hilt and rapidly withdraw it. Blood can be collected in a Microtainer (Fisher Scientific) containing Longmire's or Queen's buffer using a capillary tube; it can subsequently be stored in buffer or absorbed using commercial-grade filter paper (e.g., Whatman Grade 903, Ahlstrom Grade 226, or Munktell TFN), Nobuto blood filter strips (Advantec), or FTA cards and dried. Makowski et al. (1997) recommended usage of guanidium thiocyanate-impregnated filter paper (i.e., GT-903), which binds PCR inhibitors and preserves DNA in an aqueous extractable form, eliminating the need for expensive extraction kits. This modification may be particularly useful for mammal blood, which has a low DNA copy number relative to birds or other vertebrates with nucleated DNA. FTA cards are expensive but offer the following benefits: easy storage and shipping; issues associated with blood-borne pathogens are inactivated, reducing biohazard concerns; and a very small amount of blood can be used for any application (e.g., PCR, molecular diagnostics). Nobuto strips consist of cellulose fibers that can hold up to 1 mL of blood, which once dried is expected to be safe for storage at ambient temperature. All three options have the advantage of being able to be safely stored at room temperature once dry; however, to reduce the chance of mold and to avoid reducing the effectiveness of the cellulose to work effectively, care must be taken to not store prior to the blood drying and to avoid oversaturating the paper. Ideally, you should air-dry the filter papers and then store in individual envelopes or small plastic bags with silica desiccant.

Tail bleeding. Second, blood can be extracted from the caudal vein in the tail using a needle or lancet to puncture the caudal vein (Omaye et al. 1987). This method can be difficult with some species, such as very small rodents or small mammals with hairy tails. Warming the animal, particularly the base of the tail where the vein is accessed, will increase blood flow. Alternatively, excising the distal 1–2 mm of the tail can yield a small amount of blood, and the tail tip can also be used for DNA extraction. Mild anesthesia is recommended to minimize animal stress.

Retro-orbital. Third, blood can be obtained from the retro-orbital sinus. Disinfect the area with EtOH prior to bleeding and apply pressure or a cauterizing agent (silver nitrate or Kwik Stop) to stop bleeding. Retro-orbital bleeding is more invasive than other techniques and should be performed by trained and experienced persons (see detailed method in Mills et al. 1995, Hoff 2000). The use of very short-acting anesthesia (e.g., isoflurane or sevo-

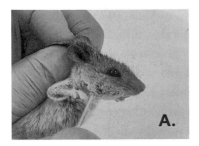  

Fig. 4.6. Submandibular blood collection from a *Mastomys natalensis*. A Goldenrod animal lancet is used to puncture the vascular bundle (*left*), which will produce a blood drop (*center*) that can be absorbed using a Nobuto strip (*right*). (Photo: Bonginkosi Charles Gumbi)

flurane) in a plastic bag will immobilize rodents in 15–20 seconds, thereby making the procedure safer for the rodent and the handler.

While all of the above techniques are commonly used, Golde et al. (2005) advocated the use of the first method, which has fewer disadvantages than the latter two. For example, bleeding from the facial vein using a lancet can be performed repeatedly on the same individual without adverse effect, whereas it is inhumane to repeatedly remove sections of the tail tip of one animal. Bleeding from the facial vein with a lancet can be performed easily after practice and requires no anesthesia, whereas bleeding from the retro-orbital sinus requires both anesthesia and training. However, Douglass et al. (2000) have shown in field studies of several rodent species that retro-orbital bleeding without anesthesia did not result in any immediate increase in mortality.

Alternative source of blood. Depending on the animal's size and other factors, alternative bleeding methods may be optimal for different groups of rodents. For example, in 115–340 g bushy-tailed woodrats (*Neotoma cinerea*), with the use of a tourniquet around the thigh, a 24-gauge needle was used to puncture the femoral vein; this was followed by blood collection in heparinized microcapillary tubes. In 2–5 kg yellow-bellied marmots (*Marmota flaviventris*), also with a tourniquet, 2 mL blood was drawn with a 21-gauge needle into a Vacutainer (Frase and Van Vuren 1989). In both the woodrats and marmots, before bleeding, individuals were sedated using an intramuscular injection of ketamine hydrochloride, which was found to have no long-term adverse effects on the health or survival of individuals. A method of bleeding rodents that has been applied previously but is generally considered inhumane (and banned in some countries) involves cardiac puncture (drawing blood directly from the heart using a needle and syringe), but this is not recommended (Golde et al. 2005) unless the animal is being sacrificed. However, cardiac puncture is arguably the best method for obtaining large volume of blood. Hoff (2000) provides a comprehensive review of bleeding methods in the laboratory mouse (*Mus musculus domesticus*). Parasuraman et al. (2010) provide a comprehensive overview of blood sample collection

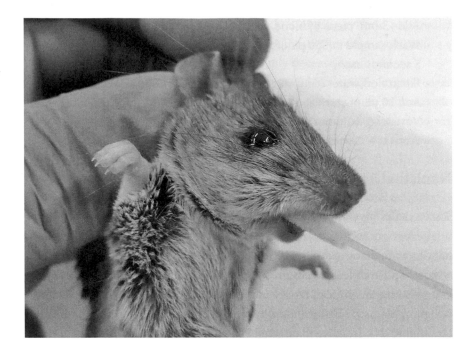

Fig. 4.7. Swab being used to collect cheek cells from a *Mastomys natalensis (Photo: Bong-inkosi Charles Gumbi)*

prominent tail vertebrae. Species temperament (e.g., those that tend to bite) may be a limitation for buccal swabs in some cases, increasing handling time and potential injury to the handler and animal. Swabs should be stored in a Tris(hydroxymethyl)aminomethane hydrochloride/EDTA (Tris-HCl/EDTA) buffer and frozen if not extracted within a short amount of time. Importantly, oral swabbing will capture DNA from a broad range of environmental DNA, including bacteria and viruses (Zheng et al. 2016), making it less ideal for applications where only DNA from targeted species is required but where PCR primers or species-specific probes are not available.

Epithelial cells can be easily obtained from more docile animals, though in our experience restraining animals for enough time to be sure to obtain sufficient cells tends to require more time (and thus more stress) than ear clipping or hair pulling. Hand-restrained laboratory mice exposed to buccal swabbing, hair pulling, tail biopsy, and ear punching were shown to produce very similar short-term (1 hour) physiological responses (increased heart rate and body temperature), suggesting that the restraint step may be of critical importance, more so than the perceived invasiveness of the sampling procedure (Cinelli et al. 2007).

Feces

Fecal pellets can often be collected directly from the animal during handling or from the capture bag (see earlier). For smaller mammals (10–20 g) at least three pellets will be needed for adequate DNA for most genotyping procedures. Fecal pellets can also be retained for dietary analysis of small mammals using either macroscopic or molecular methods (see Chapter 11), and for studying cortisol (stress hormone) levels. Care must be taken if feces sampling is for the purposes of dietary analyses or to study hormones. Small mammals captured with baited traps may digest and defecate bait within a few hours of being in at trap, contaminating feces with natural dietary items with feces derived from bait. To avoid this, passive collection of feces (Chapter 2) or trapping for very short periods of time, trapping without bait, or trapping with scent bait could be used. Pellets can be stored in EtOH, silica, or Drierite desiccant in microcentrifuge tubes.

Hair

Hair has been useful as a source of DNA, particularly via hair snares on larger mammals, but hair is also useful for examining stress hormones and for stable isotope analyses. The uses of stable isotope analysis in mammalian research has been reviewed in detail in a special issue of the *Journal of Mammalogy* (Ben-David and Flaherty 2012).

How hair is collected will depend on the end use. For DNA purposes, the hair follicle is needed. Tufts of hair can be easily plucked from the neck line or ventral body with little effort needed to dislodge them. Remove only as many as needed using anatomical forceps, without leaving visible bald patches (Schmitteckert et al. 1999). The hair follicles can be air-dried in plastic sandwich bags or coin envelopes or preserved in EtOH. This method does not require anesthesia. In a case where hair samples are collected for radio-isotope or stress-hormone analyses, the hair follicle is not necessary, and hair samples can be cut with a scissors close to the root or shaved. Hair samples from museum voucher specimens can also be used for stable isotope analysis.

Toe Clips

The removal of one or more phalange provides a permanent mark (Chapter 5) as well as tissue adequate for molecular studies. Although once a common marking technique, removing phalanges now requires careful consideration of the species' ecology and how toe removal may affect digging or arboreal behavior (Sikes and the Animal Care and Use Committee of the American Society of Mammalogists 2016).

Tail Tips

Although tails tips have been common sources for tissues, they contain a variety of tissues including cartilage, nerves, and bone, and thus this procedure has declined due to perceived stress on the animal (Hankenson et al. 2008).

century mammalogy." *Journal of Mammalogy* 100: 733–750. https://doi.org/10.1093/jmammal/gyy148.

Corthals, A., A. Martin, O. M. Warsi, M. Woller-Skar, W. Lancaster, A. Russel, et al. 2015. "From the field to the lab: Best practices for field preservation of bat specimens for molecular analysis." *PLoS ONE* 10: e0118994.

Dessauer, H. C., R. A. Menzies, and D. E. Fairbrothers. 1984. "Procedures for collecting and preserving tissues for molecular studies." In *Collections of Frozen Tissues: Value, Management, Field and Laboratory Procedures, and Directory of Existing Collections*, edited by H. C. Dessauer and M. S. Hafner, 21–24. Lawrence: Association of Systematics Collections, University of Kansas Press.

Dillon, N., A. D. Austin, and E. Bartowsky. 1996. "Comparison of preservation techniques for DNA extraction from hymenopterous insects." *Insect Molecular Biology* 5: 21–24.

Douglass, R. J., A. J. Kuenzi, T. Wilson, and R. C. Van Horne. 2000. "Effects of bleeding nonanesthetized wild rodents on handling mortality and subsequent recapture." *Journal of Wildlife Diseases* 36: 700–704.

Dunnum, J., J. Malaney, and J. Cook. 2020. "Sustained impact of holistic specimens for mammalogy and parasitology in South America: Sydney Andersons's legacy." *Therya* 11: 347–358. https://doi.org/10.12933/therya-20-1011.

Dunnum, J. L., R. Yanagihara, K. M. Johnson, B. Armien, N. Batsaikhan, L. Morgan, and J. A. Cook. 2017. "Biospecimen repositories and integrated databases as critical infrastructure for pathogen discovery and pathobiology research." *PLoS Neglected Tropical Diseases* 11: e0005133. https://doi.org/10.1371/journal.pntd.0005133.

Ferreira, C. M., H. Sabino-Marques, S. Barbosa, P. Costa, C. Encarnação, R. Alpizar-Jara, et al. 2018. Genetic non-invasive sampling (gNIS) as a cost-effective tool for monitoring elusive small mammals. *European Journal of Wildlife Research* 64: 46. https://doi.org/10.1007/s10344-018-1188-8.

Frase, B. A., and D. Van Vuren. 1989. "Techniques for immobilizing and bleeding marmots and woodrats." *Journal of Wildlife Disease* 25: 444–445.

Gagea-Iurascu, M., and S. Craig. 2012. "Euthanasia and necropsy." In *The Laboratory Rabbit, Guinea Pig, Hamster, and Other Rodents*, American College of Laboratory Animal Medicine Series, edited by M. A. Suckow, K. A. Stevens, and R. P. Wilson, 117–139. Amsterdam: Academic Press.

Ge, X. Y., J. L. Li, X. L. Yang, A. A. Chmura, G. Zhu, J. H. Epstein, et al. 2013. "Isolation and characterization of a bat SARS-like coronavirus that uses the ACE2 receptor." *Nature* 503: 535–538.

Golde, W. T., P. Gollobin, L. L. Rodriguez. 2005. "A rapid, simple method for submandibular bleeding of mice using a lancet." *Lab Animal* 34: 39–43.

Grabherr, M. G., B. J. Haas, M. Yassour, J. Z. Levin, D. A. Thompson, I. Amit, et al. 2011. "Trinity: Reconstructing a full length transcriptome without a genome from RNA-Seq data." *Nature Biotechnology* 29: 644–652.

Greer, C. E., J. K. Lund, and M. M. Manos. 1991. "PCR amplification from paraffin embedded tissue: Recommendations on fixatives for long-term storage and prospective studies." *PCR Methods and Applications* 1: 46–50.

Griffin, M., and H. Kolberg. 2004. *Preserving Biological Specimens: A Guide for Namibians*. Windhoek, Namibia: Biosystematics Working Group, National Biodiversity Programme.

Hall, E. R. 1962. "Collecting and preparing study specimens of vertebrates." *University of Kansas Museum of Natural History, Miscellaneous Publications* 30: 1–46.

Han, B. A., J. P. Schmidt, S. E. Bowden, and J. M. Drake. 2015. "Rodent reservoirs of future zoonotic diseases." *Proceedings of the National Academy of Sciences, USA* 112: 7039–7044.

Hankenson, F. C., L. M. Garzel, D. D. Fischer, B. Nolan, and K. D. Hankenson. 2008. "Evaluation of tail biopsy collection in laboratory mice (*Mus musculus*): Vertebral ossification, DANN quantity, and acute behavioral responses." *Journal of the American Association for Laboratory Animal Science* 47: 10–18.

Hanner, R., A. Corthals, and H.C. Dessauer. 2005. "Salvage of genetically valuable tissues following a freezer failure." *Molecular Phylogenetics and Evolution* 34: 452–455.

Herbreteau, V., S. Jittapalapong, W. Rerkamnuaychoke, Y. Chaval, J. F. Cosson, and S. Morand, eds. 2011. *Protocols for Field and Laboratory Rodent Studies*. Bangkok: Kasetsart University Press.

Hoff, J. 2000. "Methods of blood collection in the mouse." *Lab Animal* 29: 47–53.

Holson, R. R. 1992. "Euthanasia by decapitation: Evidence that this technique produces prompt, painless unconsciousness in laboratory rodents." *Neurotoxicology and Teratology* 14: 253–257.

ICAO Dangerous Goods Panel. 2009. "Transport of Museum Specimens." https://www.icao.int/safety/DangerousGoods/Working%20Group%20of%20the%2Whole%2009/DGPWG.09.WP.061.2.en.pdf.

International Air Transport Association (IATA). 2012. *Dangerous Goods Regulations*. Montreal: International Air Traffic Association.

Ivanova, N. V., E. L. Clare, and A.V. Borisenko. 2012. "DNA barcoding in mammals." In *DNA Barcodes: Methods and Protocols*, edited by W. J. Kress and D. L. Erickson, 153–182. Totowa, NJ: Humana Press.

Jones, C. P., S. Carver, and L. V. Kendall. 2012. "Evaluation of common anesthetic and analgesic techniques for tail biopsy in mice." *Journal of the American Association for Laboratory Animal Science* 51: 808–814.

Kassel, R., and S. Levitan. 1953. "A jugular technique for the repeated bleeding of small animals." *Science* 118: 563–564.

Kawasaki, E. S. 1990. "Sample preparation from blood, cells, and other fluids." In *PCR Protocols: A Guide to Methods and Applications*, edited by M. A. Innis, D. H. Gelfand, J. J. Sninsky, and T. J. White. San Diego: Academic Press.

Kilpatrick, C. W. 2002. "Noncryogenic preservation of mammalian tissues for DNA extraction: An assessment of storage methods." *Biochemical Genetics* 40: 53–62.

Koprowski, J. L. 2002. "Handling tree squirrels with a safe and efficient restraint." *Wildlife Society Bulletin* 30: 101–103.

Leary, S., W. Underwood, R. Anthony, S. Cartner, D. Corey, T. Grandin, et al. 2013. *AVMA Guidelines for the Euthanasia of Animals*. Schaumburg, IL: American Veterinary Medical Association.

Lemmon, A. R., S. A. Emme, and E. Moriarty Lemmon. 2012. "Anchored hybrid enrichment for massively high-throughput phylogenomics." *Systematic Biology* 61: 727–744.

Longmire, J. L., M. Maltbie, and R. J. Baker. 1997. "Use of 'lysis buffer' in DNA isolation and its implications for museum collections." *Museum of Texas Tech University* 163: 1–3.

Luis, A. D., D. T. Hayman, T. J. O'Shea, P. M. Cryan, J. R. Pulliam, J. N. Mills, et al. 2013. "A comparison of bats and rodents as reservoirs of zoonotic viruses: Are bats special?" *Proceedings of the Royal Society of London B* 280: 20122753.

Makowski, G. S., E. L. Davis, and S. M. Hopfer. 1997. "Amplification of Guthrie card DNA: Effect of guanidine thiocyanate on binding of natural whole-blood PCR inhibitors." *Journal of Clinical Laboratory Analysis* 11: 87–93.

Malaney, J. L., and J. A. Cook. 2018. "A perfect storm for mammalogy: Declining sample availability in a period of rapid environmental degradation." *Journal of Mammalogy* 99: 773–788. https://doi.org/10.1093/jmammal/gyy082.

Martin, R. E., R. H. Pine, and A. F. DeBlase. 2001. *A Manual of Mammalogy: With Keys to Families of the World*. 3rd ed. Long Grove, IL: Waveland Press.

Massie, H., H. Pfitzinger, and P. Mangin. 1972. "The kinetics of degradation of DNA and RNA by H_2O_2." *Biochimica et Biophysica Acta* 272: 539–548.

McCabe, E. R. 1991. "Utility of PCR for DNA analysis from dried blood spots on filter paper blotters." *PCR Methods and Applications* 1: 99–106.

McCleery, R. A., R. R. Lopez, and N. J. Silvy. 2007. "An improved method for handling squirrels and similar-size mammals." *Wildlife Biology in Practice* 3 (1): 39–42.

Menke, S., M. A. F. Gillingham, K. Wilhelm, and S. Sommer. 2017. "Home-made cost effective preservation buffer is a better alternative to commercial preservation methods for microbiome research." *Frontiers in Microbiology* 8: 102

Michaud, C. L., and D. R. Foran. 2011. "Simplified field preservation of tissues for subsequent DNA analysis." *Journal of Forensic Sciences* 56: 846–852.

Mills, J. N., J. E. Childs, T. G. Ksiazek, and C. J. Peters. 1995. *Methods for Trapping and Sampling Small Mammals for Virologic Testing*. Atlanta GA: US Department of Health & Human Services. https://stacks.cdc.gov/view/cdc/11507.

Mills, J. N., T. L. Yates, J. E. Childs, R. R. Parmenter, T. G. Ksiazek, P. E. Rollin, et al. 1995. "Guidelines for working with rodents potentially infected with hantavirus." *Journal of Mammalogy* 76: 716–722.

Morgan, K. T. 1997. "Brief review of formaldehyde carcinogenesis in relation to rat nasal pathology and human health risk assessment." *Toxicologic Pathology* 25: 291–307.

Morse, S. S, J. A. K. Mazet, M. Woolhouse, C. R. Parrish, D. Carroll, W. B. Karesh, et al. 2012. "Prediction and prevention of the next pandemic zoonosis." *The Lancet* 380: 1956–1965.

Murphy M. A., L. P. Waits, K. C. Kendall, S. K. Wasser, J. A. Higbee, and R. Bogden. 2002. "An evaluation of long-term preservation methods for brown bear (*Ursus arctos*) faecal DNA samples." *Conservation Genetics* 3: 435–440.

Nagorsen, D. W., and Peterson, R. L. 1980. *Mammal Collectors' Manual: A Guide for Collecting, Documenting, and Preparing Mammal Specimens for Scientific Research*. Toronto: Royal Ontario Museum.

National Research Council (NRC). 2011. *Guide for the Care and Use of Laboratory Animals*. Washington, DC: National Academies Press.

Olival, K. J., P. R. Hosseini, C. Zambrana-Torrelio, N. Ross, T. L. Bogich, and P. Daszak. 2017. "Host and viral traits predict zoonotic spillover from mammals." *Nature* 546: 646–650.

Omaye, S. T., J. H. Skala, M. D. Gretz, E. E. Schaus, and C. E. Wade. 1987. "Simple method for bleeding the unanesthetized rat by tail venipuncture." *Laboratory Animals* 21: 261–264.

Panasci, M., W. B. Ballard, S. Breck, D. Rodriguez, L. D. Densmore III, D. B. Wester, et al. 2011. "Evaluation of fecal DNA preservation techniques and effects of sample age and diet on genotyping success." *The Journal of Wildlife Management* 75: 1616–1624.

Parasuraman, S., R. Raveendran, and R. Kesavan. 2010. "Blood sample collection in small laboratory animals." *Journal of Pharmacology and Pharmacotherapeutics* 1: 87–93.

Parker, W. T., L. I. Muller, R. R. Gerhardt, D. P. O'Rourke, and E. C. Ramsay. 2008. "Field use of isoflurane for safe squirrel and woodrat anesthesia." *The Journal of Wildlife Management* 72: 1262–1266.

Pyke, G. H., and P. R. Ehrlich. 2010. "Biological collections and ecological/environmental research: A review and some observations and a look into the future." *Biological Reviews* 85: 247–266.

Rammler, D. H. 1971. "Use of DMSO in enzyme-catalyzed reactions." In *Dimethyl Sulfoxide*, vol. 1, edited by S. W. Jacob, E. E. Rosenbaum, and D. C. Wood, 189–206. New York: Marcel-Dekker.

Renshaw, M. A., B. P. Olds, C. L. Jerde, M. M. McVeigh, and D. M. Lodge. 2015. "The room temperature preservation of filtered environmental DNA samples and assimilation into a phenol-chloroform-isoamyl alcohol DNA extraction." *Molecular Ecology Resources* 15: 168–176.

Rocha, L. A., A. Aleixo, G. Allen, F. Almeda, C. C. Baldwin, M. V. L. Barcla, et al. 2014. "Specimen collection: An essential tool." *Science* 344: 814–815.

Schmitteckert, E. M., C. M. Prokop, and H. J. Hedrich. 1999. "DNA detection in hair of transgenic mice—a simple technique minimizing the distress on the animals." *Laboratory Animals* 33: 385–389.

Setzer, H. W. 1968. "Directions for preserving mammals for museum study." *Smithsonian Institution Information Leaflet* 380: 1–19.

Seutin, G., B. N. White, and P. T. Boag. 1991. "Preservation of avian blood and tissue samples for DNA analysis." *Canadian Journal of Zoology* 69: 82–90.

Sikes, R., and the Animal Care and Use Committee of the American Society of Mammalogists. 2016. "2016 Guidelines of the American Society of Mammalogists for the use of wild mammals in research and education." *Journal of Mammalogy* 97: 663–688. https://doi.org/10.1093/jmammal/gyw078.

Suarez, A. V., and N. D. Tsutsui. 2004. "The value of museum collections for research and society." *BioScience* 54: 66–74.

Taberlet, P., L. P. Waits, and G. Luikart. 1999. "Noninvasive genetic sampling: Look before you leap." *Trends in Ecology and Evolution* 14: 323–327.

Taylor, L. H., S. M. Latham, and M. E. Woolhouse. 2001. "Risk factors for human disease emergence." *Philosophical Transactions of the Royal Society, London Biological Sciences* 356: 983–989.

Williams, S. T. 2007. "Safe and legal shipping of tissue samples: Does it affect DNA quality?" *Journal of Molluscan Studies* 73: 416–418.

Wilson, D. E., F. R. Cole, J. D. Nichols, R. Rudran, and M. S. Foster (eds.). 1996. *Measuring and Monitoring Biological Diversity: Standard Methods for Mammals*. Washington, DC: Smithsonian Institution Press.

Yang, C. W. 2018. "Leptospirosis renal disease: Emerging culprit of chronic kidney disease unknown etiology." *Nephron* 138: 129–136.

tag at the same time, so the procedure may require two handlers. Even with experienced handlers, there is always a risk of ripping the skin or injuring the animal, especially when handing smaller rodents (<20 g). Researchers can also be deterred by the cost of a reader and pit tags (≈ $2.00 per tag), which can add up if you are catching thousands of individuals (Jung et al. 2019). Another drawback to the use of pit tags is that different manufacturers may use different radio frequencies, meaning that a given reader is unlikely to be able to read every tag (Gibbons and Andrews 2004). Probably the greatest concern with PIT tags is that tags can fail, be rejected by the animals, or migrate around the animal's body, all of which can hinder the identification of marked individuals (Gibbons and Andrews 2004). Most small mammals appear to have relatively high retention rates (92%–98%, Table 5.1), and tag losses usually occur within a week of tagging (Fokidis et al. 2006, Lebl and Ruf 2010), but other small mammals, such as fox squirrels (*Sciurus niger*), pocket gophers (Geomyidae), and deer mice (*Peromyscus maniculatus*), have considerably lower retention rates (Table 5.1). One proven and commonly used practice for reducing tag loss and improving retention is to use a surgical or tissue adhesive on the injection site (Lebl and Ruf 2010).

Table 5.1. Comparison of retention of ear tags and passive integrated transponder (PIT) tags for different species of small mammal by study. Rates presented as the percent of retained tags with the overall number of tagged animals and number of tags lost or not functioning in brackets

Species	Ear	PIT	Source
Microtus ochrogaster	88 (502/58)	95 (502/24)	Harper and Batzli 1996
Microtus ochrogaster	98 (325/7)		Krebs et al. 1969
Microtus ochrogaster	84 (191/31)		Wood and Slade 1990
Microtus pennsylvanicus	91 (339/30)	95 (339/16)	Harper and Batzli 1996
Microtus pennsylvanicus	95 (492/25)		Krebs et al. 1969
Glaucomys volans	81 (2277/435)	92 (2277/191)	Fokidis et al. 2006
Mus musculus	97 (124/4)	94 (124/7)	Fokidis et al. 2006
Peromyscus spp.	98 (374/8)	96 (374/16)	Fokidis et al. 2006
Sigmodon hispidus	96 (112/5)	96 (112/4)	Fokidis et al. 2006
Peromyscus maniculatus	92 (110/9)	68 (110/35)	Kuenzi et al. 2005
Dipodomys ingens	87 (1439/191)	97 (1278/38)	Williams et al. 1997
Dipodomys heermanni	87 (306/39)	97 (306/8)	Williams et al. 1997
Dipodomys nitratoides	75 (904/225)	93 (865/63)	Williams et al. 1997
Spermophilus townsendii		97 (737/25)	Schooley et al. 1993
Sciurus niger		83*	Silvy et al. 2012
Geomys spp.		8 (13/12)	Silvy et al. 2012
Glis glis		98 (1390/28)	Lebl and Ruf 2010

*Sample size not given.

Other Attachments and Markers

There are a myriad of methods, other than ear tags and PIT tags, that have been used to mark small mammals (see Twigg 1975). Here, we touch on several less commonly used markers that are still utilized in research. Ring banding is ubiquitously practiced for birds, and the placement of a numbered metal ring band above the ankle was one of the earliest ways of identifying individual small mammals (Chitty 1937). However, there are a number of problems with this technique for small mammals. Bands that are placed too tightly do not give the animals room to grow and can restrict circulation causing swelling, infection, loss of limbs, and even death (Twigg 1975). Alternatively, loose bands may slip off or damage feet (Fullagar and Jewell 1965). Additionally, leg bands are highly susceptible to gnawing from rodents. For these reasons, ringing the legs of a small mammal is rarely practiced (Twigg 1975), but it has also not been completely abandoned (Meserve et al. 1996, Yunger et al. 2002, Fernandes et al. 2012). Like leg bands, colored streamers are more common on birds (and large mammals), but they have also been attached to their ears of small mammals to observe behaviors at a distance (Ritchie 1988). However, this methodology is probably of limited use for most small mammals because of the potential increased predation risk resulting from increased visibility.

Radiotelemetry and GPS telemetry tags can be used to identify small mammals and follow their movements (see Chapter 8), but collars used to attach these tags have been used as long-term markers in their own right. For example, collars have been used as identifiers for squirrels (*Sciurus* spp.) where they can be deployed remotely and can last longer than ear tags (Downing and Marshall 1959, Wood 1976, Mahan et al. 1994). Nylon cable ties with tags, colored tape, or beads have also been used as a quick and cheap method for marking degus (*Octodon degus*), rats (*Rattus* spp.), African ice rats (*Otomys sloggetti*), and other small mammals (Berdoy et al. 1995, Ebensperger and Hurtado 2005, Pillay and Rymer 2017). Metal necklaces with identification tags, commonly used on bats, have also occasionally been used as identifiers of terrestrial small mammals (Madinah et al. 2011). Collars or necklaces are likely to be more effective on small mammals with a distinct transition from neck to head, otherwise collars are likely to slip off. A collar should be tight enough that it does not move up and down the neck and cause sores or abrasions but loose enough as to not cut off blood circulation (Silvy et al. 2012). Foam rubber, elastic, and degradable stiches can be integrated into collar designs to allow collars to expand with the growth of the animal (Byrom 1997, Smith et al. 1998). Regardless of how well the collar is fitted, it is still vulnerable to loss from chewing and grooming from the individual and its conspecifics (Mahan et al. 1994).

use in both genetic tagging and PIT tags and a decrease in the use of toe clipping (Jung et al. 2019). Additionally, small mammal researchers believe there is a clear need for more formal assessments of the cost and benefits of different marking techniques so they can make informed decisions for their research (Jung et al. 2019).

REFERENCES

Ambrose, H. W., III. 1972. "Effect of habitat familiarity and toe clipping on rate of owl predation in *Microtus pennsylvanicus*." *Journal of Mammalogy* 53 (4): 909–912.

Armitage, K. B., and K. S. Harris. 1982. "Spatial patterning in sympatric populations of fox and gray squirrels." *American Midland Naturalist* 108 (2): 389–397. https://doi.org/10.2307/2425501.

Baird, P., D. Robinette, and S. A. Hink. 2017. "A remote marking device and newly developed permanent dyes for wildlife research." *Wildlife Society Bulletin* 41 (4): 785–795.

Berdoy, M., P. Smith, and D. W. Macdonald. 1995. "Stability of social status in wild rats: Age and the role of settled dominance." *Behaviour* 132 (3–4): 193–212.

Bergstedt, B. 1966. "Home ranges and movements of the rodent species *Clethrionomys glareolus* (Schreber), *Apodemus flavicollis* (Melchior) and *Apodemus sylvaticus* (Linné) in southern Sweden." *Oikos* 17: 150–157. https://doi.org/10.2307/3564939.

Bissonette, J. A., and S. A. Rosa. 2009. "Road zone effects in small-mammal communities." *Ecology and Society* 14 (1): 27.

Blair, W. F. 1941. "Techniques for the study of mammal populations." *Journal of Mammalogy* 22 (2): 148–157. https://doi.org/10.2307/1374909.

Borremans, B., V. Sluydts, R. H. Makundi, and H. Leirs. 2015. "Evaluation of short-, mid- and long-term effects of toe clipping on a wild rodent." *Wildlife Research* 42: 143–148.

Braude, S., and D. Ciszek. 1998. "Survival of naked mole-rats marked by implantable transponders and toe-clipping." *Journal of Mammalogy* 79 (1): 360–363. https://doi.org/10.2307/1382873.

Brouard, M. J., T. Coulson, C. Newman, D. W. Macdonald, and C. D. Buesching. 2015. "Analysis on population level reveals trappability of wild rodents is determined by previous trap occupant." *PLoS ONE* 10 (12). https://doi.org/10.1371/journal.pone.0145006.

Byrom, A. E. 1997. "Population ecology of arctic ground squirrels in the boreal forest during the decline and low phases of a snowshoe hare cycle." PhD diss., University of British Columbia.

Chitty, D. 1937. "A ringing technique for small mammals." *Journal of Animal Ecology* 6: 36–53. https://doi.org/10.2307/1057.

Downes, S. J., K. A. Handasyde, and M. A. Elgar. 1997. "The use of corridors by mammals in fragmented Australian eucalypt forests." *Conservation Biology* 11 (3): 718–726. https://doi.org/10.1046/j.1523-1739.1997.96094.x.

Downing, R. L., and C. M. Marshall. 1959. "A new plastic tape marker for birds and mammals." *Journal of Wildlife Management* 23 (2): 223–224. https://doi.org/10.2307/3797646.

Ebensperger, L. A., and M. J. Hurtado. 2005. "On the relationship between herbaceous cover and vigilance activity of degus (*Octodon degus*)." *Ethology* 111 (6): 593–608. https://doi.org/10.1111/j.1439-0310.2005.01084.x.

Evans, F. C., and R. Holdenried. 1943. "A population study of the Beechey ground squirrel in central California." *Journal of Mammalogy* 24 (2): 231–260. https://doi.org /10.2307/1374808.

Farentinos, R.C. 1972. "Observations on ecology of tassel-eared squirrel." *Journal of Wildlife Management* 36 (4): 1234–1239. https://doi.org/10.2307/3799253.

Fernandes, F. R., L. D. Cruz, and A. X. Linhares. 2012. "Effects of sex and locality on the abundance of lice on the wild rodent *Oligoryzomys nigripes*." *Parasitology Research* 111 (4): 1701–1706. https://doi.org/10.1007/s00436-012-3009-4.

Firth, R. S. C., B. W. Brook, J. C. Z. Woinarski, and D. A. Fordham. 2010. "Decline and likely extinction of a northern Australian native rodent, the brush-tailed rabbit-rat *Conilurus penicillatus*." *Biological Conservation* 143 (5): 1193–1201. https://doi.org /10.1016/j.biocon.2010.02.027.

Fitzwater, W. D. 1943. "Color marking of mammals, with special reference to squirrels." *Journal of Wildlife Management* 7 (2): 190–192. https://doi.org/10.2307/3795724.

Fokidis, H. B., C. Robertson, and T. S. Risch. 2006. "Keeping tabs: Are redundant marking systems needed for rodents?" *Wildlife Society Bulletin* 34 (3): 764–771. https:// doi.org/10.2193/0091-7648(2006)34[764:ktarms]2.0.co;2.

Fourie, L. J., and M. R. Perrin. 1986. "Effective methods of trapping and marking rock hyrax." *South African Journal of Wildlife Research* 16 (2): 58–61.

Fullagar, P. J., and P. A. Jewell. 1965. "Marking small rodents and the difficulties of using leg rings." *Proceedings of the Zoological Society of London* 147 (2): 224–228.

Gerber, B. D., and R. R. Parmenter. 2015. "Spatial capture-recapture model performance with known small-mammal densities." *Ecological Applications* 25 (3): 695–705. https://doi.org/10.1890/14-0960.1.

Gibbons, J. W., and K. M. Andrews. 2004. "PIT tagging: Simple technology at its best." *Bioscience* 54 (5): 447–454. https://doi.org/10.1641/0006-3568(2004)054[0447:ptstai] 2.0.co;2.

Goodman, S. M., and D. Rakotondravony. 2000. "The effects of forest fragmentation and isolation on insectivorous small mammals (Lipotyphla) on the Central High Plateau of Madagascar." *Journal of Zoology* 250: 193–200. https://doi.org/10.1017 s0952836900002041.

Guidelines for the Use of Animals. 2019. "Guidelines for the treatment of animals in behavioural research and teaching." *Animal Behaviour* 147: I-X. https://doi.org /10.1016/j.anbehav.2018.12.015.

Hadow, H. H. 1972. "Freeze-branding: Permanent marking technique for pigmented mammals." *Journal of Wildlife Management* 36 (2): 645–649. https://doi.org/10.2307 /3799102.

Harper, S. J., and G. O. Batzli. 1996. "Monitoring use of runways by voles with passive integrated transponders." *Journal of Mammalogy* 77 (2): 364–369. https://doi.org /10.2307/1382809.

Hurst, J. L. 1988. "A system for the individual recognition of small rodents at a distance, used in free-living and enclosed populations of house mice." *Journal of Zoology* 215: 363–367. https://doi.org/10.1111/j.1469-7998.1988.tb04903.x.

Johnson, W. C. 2001. "A new individual marking technique: Positional hair clipping." *Southwestern Naturalist* 46 (1): 126–129. https://doi.org/10.2307/3672389.

but usually requires sacrificing the animal (Ferguson-Smith and Trifonov 2007). In fact, bone marrow has frequently been used for karyotype analysis of small mammals (Lee and Elder 1980, Dobigny et al. 2005). Molecular sequencing is more expensive but has become increasingly popular and can be done without sacrificing the animal; typically, a drop of blood, cheek swab, ear snip, or hair sample is used (Chapter 4). The analytical approach used may also vary depending on the species group (Chapter 15). For example, the barcoding gene (CO1) has proved useful for species identification of mammals (Hajibabaei et al. 2007, Meusnier et al. 2008), as has the cytochrome-b gene (Bradley and Baker 2001).

It is a good practice to photograph captured animals. This has become common practice for the taking of museum voucher specimens, where the animal can be photographed from different angles after it has been killed (Fig. 6.1). The main purpose of this is to keep digital records of the animal's original appearance, particularly its fur color and texture (the former often fading or changing in museum collections). The shape of the animal (e.g., proportions of the snout) may also change if a dry skin is produced because the skull and other internal structures will have to be removed. Photographing live animals is more difficult and usually requires some form of a cage (preferably with at least one glass side) in which the animal is placed. Such photographs show the animal in a natural pose and may make nice pictures for future talks and guidebooks. If the only aim is to have a record of the animal's head shape and fur color, then taking pictures of it while being held is fine and can be done as part of the processing procedure before releasing it. This is particularly important for researchers who are starting an ecological study in a new environment and are not 100% sure of their identifications. Even when researchers are sure of their identification, such images may act as a kind of "voucher" in case others need to be convinced that the species was really captured (Remsen 1995).

When collecting images, it is important to take photographs of the whole animal from various angles, e.g., side on, above (back), and below (belly) (Fig. 6.1). In addition, take close-ups of the head from different angles and of any other important features for that particular species (this may include the tail, hindfoot, claws, ears, etc.). For cryptic species, always take images along with tissue for genetic analysis because there is a chance that your images may provide the means of determining what external features can be used to distinguish between them.

A small word of caution about identification keys for small mammal species: they are often not completely reliable and may even be incorrect. This is especially true for the tropics, where few researchers may be conducting fieldwork. There is real need for the publication of updated identification keys

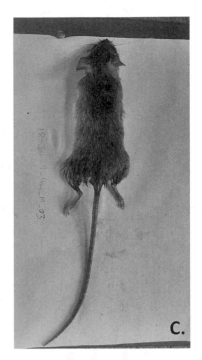

Fig. 6.1. Rodent specimen (*Aethomys ineptus*) photographed to show lateral (*A*), ventral (*B*), and dorsal (*C*) views before being deposited in a museum

in these parts of the world. If researchers cannot properly identify animals in the field, the quality of everyone's ecological work suffers.

Aging

Assessing the age of a small mammal may be useful for taxonomic purposes (e.g., subadults of one species may be confused with adults of another species) and for ecological models (e.g., calculating recruitment into a population) (see Chapter 9). Hence, depending on the objectives of the study, correctly aging (and sexing) animals can be almost as important as correctly identifying the species (Spinage 1973, Olenev 2009).

Unfortunately, there is no single tool that can be used to collectively age terrestrial small mammals. However, there are three approaches that can guide this process. The first is to look at the size of the animal, which has been done for many species, including the common African rodent *Mastomys natalensis* (Monadjem and Perrin 1998). If you know the minimum mass (or size) of adults for both males and females of the species in question, then size can be used to categorize animals as an adult or juvenile. Although this is a simple procedure in theory, requiring just a mass (or linear measurement), in reality many small mammals do not perfectly conform to such simple rules (Leirs and Verheyen 1995, Monadjem 1998). Hence, a simple size rule cannot be used with great accuracy but suffices as a useful guide.

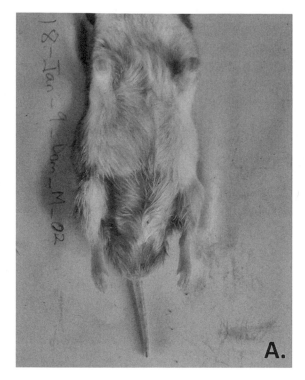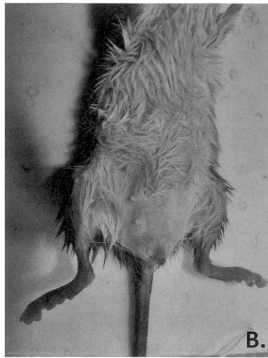

Fig. 6.2. Male rodent *Saccostomus campestris* (*A*) with enlarged and descended testes (breeding) and female rodent *Grammomys* sp. (*B*) with small nipples (nonbreeding)

A second approach is to examine the reproductive condition of the animal: males with descended testes and females with a perforate vagina and enlarged nipples are clearly ready for breeding (and hence classed as adult) (Fig. 6.2). There are two difficulties with this approach. First, not all small mammals show such clear reproductive features (see "Sexing," below). Second, it is now well established that some small mammals, such as murid rodents, can become reproductively active at typically "subadult" sizes if environmental conditions are favorable, such as a limited number of adults in the population (Bondrup-Nielsen 1986, Priotto et al. 2006).

A third approach is to examine the pelage of the animal. Juvenile and subadult small mammals typically have more uniform and softer fur than adults, but this varies considerably between species and groups. This approach is only effective if the researcher is familiar with the various juvenile/subadult coats present in the species he or she is studying.

In reality, a combination of the three approaches may be the best bet for getting around the difficult task of aging small mammals in the field. There are also a number of ways in which small mammal specimens (i.e., dead animals) can be aged (Morris 1972). One commonly used method is to examine tooth wear; this is typically based on cleaned skulls in museum collections (Perrin 1979, Chimimba and Dippenaar 1994). Tooth wear is generally used to assign the specimens to a number of tooth-wear classes (anywhere from 4 or 5 to over 12 classes), with the hope that these represent different age groups. So, young animals have the least worn teeth, and old animals the most worn teeth. However, it needs to be noted that the age of a small mammal cannot be determined with much accuracy using this method primarily because tooth wear is not just a function of age but also of diet (harder foods will wear the teeth more rapidly than softer material). Furthermore, most studies of teeth wear are based on a subjective assessment by the researcher (i.e., without any measurements being taken). The researcher simply identifies a representative skull for each tooth-wear class and then assigns the remaining skulls to one of these. Despite this, many studies have used such tooth-wear classes with considerable success (Olenev 2009).

It is, however, possible to add objectivity to the study of tooth wear by measuring the reduction in height of teeth, which has been done for some shrews (Morris 1972). Teeth can also be used to age mammals by examining the annuli in their cementum (Adams and Watkins 1967, Fogl and Mosby 1978). For example, in the California ground squirrel (*Spermophilus beecheyi*), successive light and dark rings (annuli) are laid down on the cementum of the third molar tooth, which is precise enough to age an individual up to four years old (Adams and Watkins 1967). The American beaver (*Castor canadensis*) has been aged by a combination of tooth eruption (premolars and molars) and annuli around the mandibular molars viewed in longitudinal section (Van Nostrand and Stephenson 1964). However, this requires a considerable amount of preparation in making sections of teeth, and it results in permanent damage to the specimen.

A more precise aging technique for small mammals involves the mass of the eye lens (Askaner and Hansson 1967, Morris 1972). The eye lens appears to grow continuously through the life of a small mammal, and therefore weighing it should give an accurate age for the individual. Indeed, studies of known-age individuals of many different species of rodent have shown that this relationship holds (Leirs and Verheyen 1995). Unfortunately, this technique is time-consuming and involves many independent steps, which are described in detail in Morris (1972), and it obviously results in the death of the animal (and is not particularly useful for ecological studies or studies of rare and endangered species). However, this method has been used for a wide range of small mammals (Butynski 1979, van Aarde 1985, Montgomery and Montgomery 1990, Neal 1996, Monadjem 1998, Janova et al. 2016).

Finally, the collagen fibers in the tail tendons of mice (and other small mammals such as opossums) increase in strength with age. The test for this has been termed the tail tendon break time, or TTBT, and involves measuring the resistance of tail tendon fibers to degradation by urea (Sloane et al. 2010). TTBT increases with age, but this relationship is highly variable, and

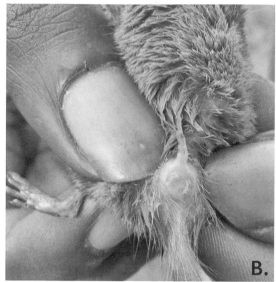

Fig. 6.3. Male shrew *Crocidura hirta* with penis hidden (*A*) and exposed (*B*). Note that the penis of shrews can be everted by gently pulling the skin on either side of the cloacal opening.

therefore, this method cannot be considered a robust biomarker of longevity (Sloane et al. 2010).

Sexing

Sexing animals is important for many reasons. Males and females may differ in size and/or in other morphological features, including fur color; knowing the sex will therefore aid in identification. Sexing the animals is also vital for the calculation of various ecological parameters such as population growth rate and fluctuation (Krebs 2014). It is not surprising, then, that sexing of a captured animal is done routinely for all small mammals.

Compared with other vertebrates, sexing mammals is generally straight-forward, especially in adults. Typically, males have a visible penis and descended testes, and females have a vagina and a series of teats. However, there are many exceptions to this rule; for example, male shrews do not have a visible penis, and their testes do not descend (Fig. 6.3), making sexing rather difficult in this group. In addition, it is often difficult to assign a sex to juvenile or subadult small mammals because their sex organs have not yet developed. In such cases, care is needed to distinguish the sexes; where separation is not possible, the researcher should categorize the animal as being of unknown sex rather than guess.

When specimens are collected, the best way to sex individuals is to open up the abdominal cavity and search for either testes or ovaries (and oviducts).

Fig. 6.4. Nonbreeding female (*A*) and male (*B*) rodents *Mastomys natalensis*; note the greater distance between the urinary papilla and anus in the male

However, this option is not suitable for most field biologists who wish to release a living animal, in which case it is worth considering the following steps.

The first step in sexing is to be aware of any anomalous groups in your study area. For example, as already mentioned, shrews do not readily exhibit any external characteristics that can be used to assign them to male or female classes. The best way to acquire this information is through a regional handbook or standard mammal guide. The next step is to be familiar with how to sex animals within each group.

Rodents form the largest group of small mammals typically captured in any field site. Adult males have descended testes but no visible penis (which is covered by the urinary papilla). Adult females have a visible vagina, and if they have bred before, their nipples will be enlarged and visible (in some species, observing this may require effort on the part of the researcher, such as blowing on the belly to part the hairs covering the teats). However, depending on the season, the majority of captured rodents may be subadults (or at least individuals that have not bred before); these animals cannot be sexed by searching for the scrotum or a vagina. Instead, they can be sexed by examining the relative gap between the anus and urinary papilla; in males this is a large gap (because the testes descend into this gap when the animal matures), and in females it is a small gap (Fig. 6.4).

Shrews form another large group that is regularly captured in field studies. Sexing shrews can be problematic and should only be done on a live animal if absolutely necessary. Male and female shrews have a single cloacal opening because the anus and urinary systems (including the reproductive system) are covered by a fold of tissue. Male shrews have a penis that can be gently massaged out of this opening (Fig. 6.3); if no such structure is manipulated out, then the animal is a female.

Small marsupials are mostly restricted to South America and Australasia but abound in places where they occur. Females typically have an abdominal pouch or marsupium within which the young develop and are nourished by 2–12 (or more) teats (Vaughan et al. 2015). However, this marsupium is missing in certain groups, such as the Caenolestidae (shrew-opossums) and Dasyuridae (dasyurids), but the teats are still visible in these groups. In males, only sexually mature marsupials have descended testes, which (unlike in placental mammals) hang anterior to the penis.

Reproductive Condition

We have already mentioned reproductive condition in the sections on aging and sexing. In practice, aging, sexing, and reproductive condition are often considered together, as they are interdependent. For example, the reproductive condition of an animal will depend on its age and sex. Male mammals are reproductively active when their testes are capable of producing viable sperm, and females when their ovaries are producing eggs. Typically, in mammals, this stage is reached when the testes are descended (males) or the vagina is perforate and the teats are enlarged (females) (Fig. 6.2), a situation commonly encountered in rodents. However, descended testes are not necessary for a male small mammal to be physiologically ready for reproduction, and some males may reach sexual maturity without descended testes. Assessing reproductive condition in other small mammal groups may be difficult to impossible, unless they are being vouchered (and the abdomen is opened up for a direct inspection of the testes or ovaries).

If the animal is being sacrificed, more information can be extracted from the specimen pertaining to its reproductive condition. For example, in addition to measuring the size of the testes, the testes themselves can be preserved to be viewed under a microscope in the lab (which can accurately address the question of whether the animal was in reproductive condition or not). For female small mammals, the same can be done with the ovaries. However, of greater interest would be to count embryos in the oviduct as a measure of pre-birth litter size (Fig. 6.5). By measuring these embryos, a researcher can age them, which in turn provides a date for breeding. Furthermore, the oviducts can be inspected for "placental scars," which are tell-tale signs of previous pregnancies. These placental scars can also be used to estimate the number of pups that a female has given birth to and, therefore, as a measure of reproductive output (Davis and Emlen 1948, Martin et al. 1976, Dracup et al. 2016).

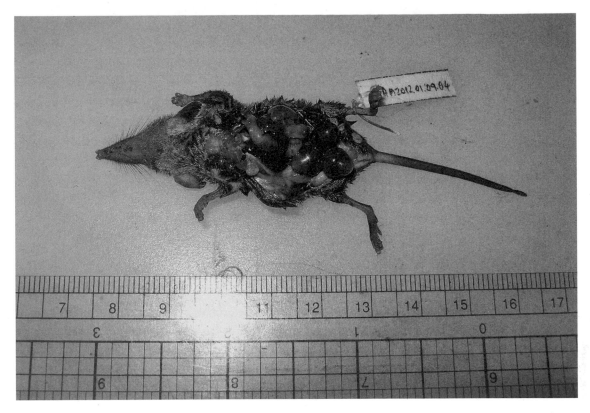

Fig. 6.5. Dissected female shrew *Crocidura* sp. specimen showing three embryos

If an animal is part of an ecological study and is not going to be sacrificed, then the relevant information that can be extracted on its reproductive condition is rather limited. For male small mammals, about the only useful information that can be recorded is whether the testes are descended or not (for those species, such as rodents, that actually develop a scrotum). For females, the situation is only slightly better. In rodents, for example, a female can be scored as "perforate vagina," indicating that it is capable of breeding. In addition, the condition of her nipples can be assessed: they can be swollen and bare (no hair immediately around them), indicating that she is lactating; they can be enlarged but not bare, showing that she has bred before; or they can be tiny and hard to see under the fur, showing that she has never bred. Finally, a pregnant female (with large fetuses) will have a bulging abdomen; the embryos (if large enough) can be felt by palpation. Be careful not to squeeze too hard, as you risk harming the developing fetuses and/or the mother.

Field Measurements

In addition to aging, sexing, and scoring the reproductive condition of captured small mammals, researchers often take measurements of the animals in the field (Hoffmann et al. 2010). For example, researchers interested in the community structure of mammals may need to know the biomass contributed by each species (Gumbi et al. 2018). Alternatively, they may be interested in daily or seasonal changes in body mass. This requires a measurement of mass for each captured animal (Grant and Birney 1979, Sollmann et al. 2015, Mason-Romo et al. 2017). Taking the mass of small mammals is almost effortless, as the animal needs to be emptied from the trap into a handling bag. Therefore, all that needs to be done is to weigh the bag with and without the small mammal. No direct handling of the animal is required, making this the easiest measurement to take. Most field ecologists routinely take the mass of captured small mammals along with age, sex, and reproductive condition.

Linear measurements are more difficult to take from living small mammals. Not only does this require extra effort from the researcher, but it probably adds extra stress to the small mammal because it will need to be handled for a greater amount of time. However, holding a small mammal for an extra few seconds to take a measurement probably does not significantly add to the stress of the animal. Before handling an animal, carefully consider which measurements you may wish to take (see "Museum Measurements," below, for a detailed description of the various measurements that you could take) so that you do not handle the animal longer than absolutely necessary.

It is worthwhile considering the various reasons why you would need to weigh or measure a living animal. The main reasons are as follows. Some small mammal species bear a close resemblance to each other but may be distinguished in the field based on size. Body length is not an easy measurement to take in the field and is liable to vary significantly not only between observers but also between different occasions by the same observer. This is because the small mammal may hunch up or stretch out, and the observer has little control over this (without hurting the animal). Tail length is easier to measure, especially with a second observer assisting. Hindfoot length is particularly useful (for identification purposes) and perhaps the easiest of the linear measurements to take (Fig. 6.6).

Museum Measurements

If an animal is sacrificed to be deposited into a museum (typically as a voucher specimen), then certain standard measurements should be taken, in line with the maxim that if an animal is killed, then ensure that as much information is extracted from that animal as possible. The following weights and external measurements are taken routinely for museum collections and are referred to as "standard museum measurements": mass, total length, head-body length, tail length, hindfoot length, and ear length (see Fig. 6.6 for how to take these

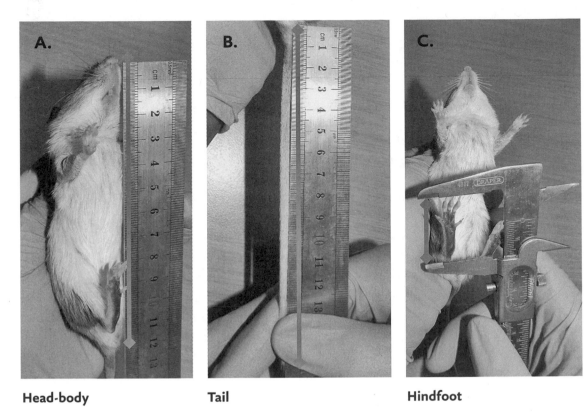

Head-body　　　　　　　**Tail**　　　　　　　**Hindfoot**

Fig. 6.6. Examples of standard museum measurements taken on a live small mammal in the field *(Photos: Bonginkosi Charles Gumbi)*

measurements). These measurements are mostly self-explanatory, but it might be worthwhile to expand on them.

"Mass" is simply the mass of the animal and for small mammals is typically taken on a reliable spring scale (e.g., Pesola) to an accuracy of 0.1 g. "Total length" is the length from the tip of the snout to the tip of the tail, with the animal stretched out without altering the structure of its body. "Head-body length" is the length from the tip of the snout to the anus, again with the animal stretched out. "Tail length" is the length of the straightened tail from the anus to the tip of the tail. Note that the tail of small mammals is often cut short (perhaps by a predator), and this should be noted under a "Comments" column so that such measurements are not used in calculating the mean tail length for the species. "Hindfoot length" can be measured in two discrete ways, and it is important not to confuse the two measurements. It can be measured with the claws, in which case it is known as hindfoot c.u. (*cum unguis*), or without the claws, where it is known as hindfoot s.u. (*sine unguis*). Neither method is superior to the other, but which method was used should be clearly stated because hindfoot c.u. is not directly comparable with hindfoot s.u. Finally, "ear length" is taken as the length from the notch inside the ear to its tip. All linear measurements in small mammals are typically reported to 0.1 mm, so you may want to use a pair of calipers for this.

Other measurements may also be taken, but if they are not standard measurements, make sure to clearly indicate how they were measured. For example, fur length may be a useful measure for a particular study, in which case, it should clearly be reported how this measurement was taken and what part(s) of the body it came from.

Tissue, Hair, Feces, and Urine Samples

You may wish to collect tissue samples for a variety of analyses, including future molecular analysis for taxonomic purposes, disease screening, or physiological/anatomical studies. Most such tissue samples can only be taken from a dead animal, and the protocol for acquiring such samples is discussed in Chapter 4. Hair, feces, and urine samples can be taken on the living animal before releasing it (Chapter 4). These samples can be extremely useful for insight into a wide range of topics, such as trophic status, diet, stress, and reproductive condition of the animal. Hair samples can be taken by simply clipping some of the animal's fur. By using stable isotope analysis (based on this hair), the trophic niche of the animal can then be determined (discussed in detail in Chapter 11). Feces samples can provide a direct view of what the animal has recently eaten. Comparing results of stable isotopes from, say, fur and feces provides insight into the general types of foods that the animal has eaten over an extended period of time. Urine samples can determine a whole host of things about the animal through the examination of various hormones, such as determining the position of female small mammals with regard to the reproductive cycle based on levels of reproductive hormones. Often hair, fecal, and urine samples can be used in combination for a more thorough understanding of an animal's physiology. For example, the animal's stress level can be assessed by examining the hormone cortisol in feces and in hair. The fecal sample provides a picture of recent stress levels, whereas the hair sample integrates stress levels over a longer period.

Conclusions and Future Directions

This chapter has summarized the various ways in which anatomical, physiological, and reproductive data can be obtained via sampling in ethically acceptable ways. The four most common pieces of information that are obtained from ecological studies generally are age, sex, reproductive condition, and mass and linear measurements. These are most easily obtained from specimens (dead animals) but can also be acquired from living animals. Other types of samples may be useful for specific studies and include feces, urine, blood, and hair. With advances in molecular, isotope, and hormonal techniques, the applicability of these samples to ecological studies is likely to increase.

REFERENCES

Adams, L., and S. G. Watkins. 1967. "Annuli in tooth cementum indicate age in California ground squirrels." *Journal of Wildlife Management* 31 (4): 836–839.

Askaner, T., and L. Hansson. 1967. "The eye lens as an age indicator in small rodents." *Oikos* 18 (1): 151–153.

Bondrup-Nielsen, S. 1986. "Investigation of spacing behaviour of *Clethrionomys gapperi* by experimentation." *Journal of Animal Ecology* 55 (1): 269–279.

Bradley, R. D., and R. J. Baker. 2001. "A test of the genetic species concept: Cytochrome-*b* sequences and mammals." *Journal of Mammalogy* 82 (4): 960–973. https://doi.org /10.1644/1545-1542(2001)082<0960:ATOTGS>2.0.CO;2.

Butynski, T. M. 1979. "Reproductive ecology of the springhaas *Pedetes capensis* in Botswana." *Journal of Zoology* 189: 221–232.

Chimimba, C. T., and N. J. Dippenaar. 1994. "Non-geographic variation in *Aethomys chrysophilus* (De Winton, 1897) and *A. namaquensis* (A. Smith) (Rodentia: Muridae) from southern Africa." *South African Journal of Zoology* 29: 107–117.

Davis, D. E., and J. T. Emlen. 1948. "The placental scar as a measure of fertility in rats." *Journal of Wildlife Management* 12 (2): 162–166.

Dobigny, G., V. Aniskin, L. Granjon, R. Cornette, and V. Volobouev. 2005. "Recent radiation in west African *Taterillus* (Rodentia, Gerbillinae): The concerted role of chromosome and climatic changes." *Heredity* 95 (5): 358–368. https://doi.org/10.1038 /sj.hdy.6800730.

Dobigny, G., V. Aniskin, and V. Volobouev. 2002. "Explosive chromosome evolution and speciation in the gerbil genus *Taterillus* (Rodentia, Gerbillinae): A case of two new cryptic species." *Cytogenetic and Genome Research* 96: 117–124. https://doi.org /10.1159/000063044.

Dobigny, G., L. Granjon, V. Aniskin, K. Ba, and V. Volobouev. 2003. "A new sibling species of *Taterillus* (Muridae, Gerbillinae) from West Africa." *Mammalian Biology* 68 (5): 299–316. https://doi.org/10.1078/1616-5047-00098.

Dracup, E. C., D. M. Keppie, and G. J. Forbes. 2016. "The short-term impact of abundant fruit upon deer mouse (*Peromyscus maniculatus*), southern red-backed vole (*Myodes gapperi*), and woodland jumping mouse (*Napaeozapus insignis*) populations." *Canadian Journal of Zoology* 94: 555–563.

Ferguson-Smith, M. A., and V. Trifonov. 2007. "Mammalian karyotype evolution." *Nature Reviews Genetics* 8 (12): 950–962. https://doi.org/10.1038/nrg2199.

Fogl, J. G., and H. S. Mosby. 1978. "Aging gray squirrels by cementum annuli in razorsectioned teeth." *Journal of Wildlife Management* 42 (2): 444–448.

Gordon, D. H. 1978. "Distribution of sibling species of the *Praomys* (*Mastomys*) natalensis group in Rhodesia (Mammalia: Rodentia)." *Journal of Zoology, London* 186: 397–401.

Grant, W. E., and E. C. Birney. 1979. "Small mammal community structure in North American grasslands." *Journal of Mammalogy* 60 (1): 23–36.

Gumbi, B. C., J. T. Shapiro, T. Mahlaba, R. McCleery, D. MacFadyen, and A. Monadjem. 2018. "Assessing the impacts of domesticated versus wild ungulates on terrestrial small mammal assemblages at Telperion Nature Reserve, South Africa." *African Zoology* 53 (1): 23–29.

Hajibabaei, M., G. A. C. Singer, E. L. Clare, and P. D. N. Hebert. 2007. "Design and applicability of DNA arrays and DNA barcodes in biodiversity monitoring." *BMC Biology* 5: 24. https://doi.org/10.1186/1741-7007-5-24.

Hoffmann, A., J. Decher, F. Rovero, J. Schaer, C. Voigt, and G. Wibbelt. 2010. "Field methods and techniques for monitoring mammals." In *Manual on Field Recording Techniques and Protocols for All Taxa Biodiversity Inventories and Monitoring*, edited by J. Eymann, J. Degreef, C. Häuser, J. C. Monje, Y. Samyn, and D. VandenSpiegel, 482–529. United Kingdom: Abc Taxa.

Janova, E., M. Heroldova, and L. Cepelka. 2016. "Rodent food quality and its relation to crops and other environmental and population parameters in an agricultural landscape." *Science of the Total Environment* 562: 164–169. https://doi.org/10.1016/j.scitotenv.2016.03.165.

Krebs, C. J. 2014. *Ecological Methodology*. 3rd ed. Menlo Park, CA: Addison-Wesley Educational Publishers Inc.

Lalis, A., A. Evin, and C. Denys. 2009. "Morphological identification of sibling species: The case of west African *Mastomys* (Rodentia: Muridae) in sympatry." *Comptes Rendus Biologies* 332 (5): 480–488. https://doi.org/10.1016/j.crvi.2008.11.004.

Lee, M. R., and F. F. B. Elder. 1980. "Yeast stimulation of bone marrow mitosis for cytogenetic investigations." *Cytogenetic and Genome Research* 26: 36–40. https://doi.org/10.1159/000131419.

Leirs, H., and W. Verheyen. 1995. *Population Ecology of Mastomys natalensis (Smith, 1834). Implications for Rodent Control in Africa. A Report from the Tanzania-Belgium Joint Rodent Research Project (1986–1989)*. Belgium: Belgian Administration for Development Cooperation.

Martin, K. H., R. A. Stehn, and M. E. Richmond. 1976. "Reliability of placental scar counts in the prairie vole." *Journal of Wildlife Management* 40 (2): 264–271. https://doi.org/10.2307/3800424.

Mason-Romo, E. D., A. A. Farías, and G. Ceballos. 2017. "Two decades of climate driving the dynamics of functional and taxonomic diversity of a tropical small mammal community in western Mexico." *PLoS ONE* 12 (12): e0189104. https://doi.org/10.1371/journal.pone.0189104.

Meusnier, I., G. A. C. Singer, J. F. Landry, D. A. Hickey, P. D. N. Hebert, and M. Hajibabaei. 2008. "A universal DNA mini-barcode for biodiversity analysis." *BMC Genomics* 9: 214. https://doi.org/10.1186/1471-2164-9-214.

Michaux, J. R., S. Kinet, G. Filippucci, R. Libois, A. Besnard, and F. Catzeflis. 2001. "Molecular identification of three sympatric species of wood mice (*Apodemus sylvaticus, A. flavicollis, A. alpicola*) in western Europe (Muridae: Rodentia)." *Molecular Ecology Notes* 1: 260–263.

Modi, W. S. 1986. "Karyotypic differentiation among two sibling species pairs of New World microtine rodents." *Journal of Mammalogy* 67 (1): 159–165.

Monadjem, A. 1998. "Reproductive biology, age structure, and diet of *Mastomys natal-*

ensis (Muridae: Rodentia) in a Swaziland grassland." *Zeitschrift Für Säugetierkunde* 63: 347–356.

Monadjem, A., and M. R. Perrin. 1998. "Effects of food supplementation and fire on small mammal community structure in a Swaziland grassland." *South African Journal of Science* 94 (2).

Montgomery, S. S. J., and W. I. Montgomery. 1990. "Intrapopulation variation in the diet of the wood mouse *Apodemus sylvaticus*." *Journal of Zoology,* London 222: 641–651.

Morris, P. 1972. "A review of mammalian age determination methods." *Mammal Review* 2 (3): 69–104. https://doi.org/10.1111/j.1365-2907.1972.tb00160.x.

Neal, B. R. 1996. "Reproductive response of *Tatera leucogaster* (Rodentia) to supplemental food and 6-methoxybenzoxazolinone in Zimbabwe." *Mammalia* 60 (4): 651–666. https://doi.org/10.1111/j.1469-7998.1991.tb04777.x.

Olenev, G. V. 2009. "Determining the age of cyclomorphic rodents: Functional-ontogenetic determination, ecological aspects." *Russian Journal of Ecology* 40 (2): 93–104. https://doi.org/10.1134/S1067413609020040.

Perrin, M. R. 1979. "Ageing criteria and population age structure of co-existing populations of *Rhabdomys pumilio* and *Otomys irroratus*." *South African Journal of Wildlife Research* 9: 84–95.

Priotto, J., C. Provensal, and J. Polop. 2006. "Effect of adults on juvenile reproduction of *Calomys venustus* (Muridae: Sigmodontinae)." *Austral Ecology* 31: 859–868. https://doi.org/10.1111/j.1442-9993.2006.01649.x.

Remsen, J. V. 1995. "The importance of continued collecting of bird specimens to ornithology and bird conservation." *Bird Conservation International* 5 (2–3): 146–180. https://doi.org/10.1017/S095927090000099X.

Russo, I.-R. M., C. T. Chimimba, and P. Bloomer. 2006. "Mitochondrial DNA differentiation between two species of *Aethomys* (Rodentia: Muridae) from southern Africa." *Journal of Mammalogy* 87 (3): 545–553.

Sloane, L. B., J. T. Stout, S. N. Austad, and G. E. McClearn. 2010. Tail tendon break time: A biomarker of aging? *Journal of Gerontology* 66: 287–294.

Sollmann, R., A. M. White, B. Gardner, and P. N. Manley. 2015. "Investigating the effects of forest structure on the small mammal community in frequent-fire coniferous forests using capture-recapture models for stratified populations." *Mammalian Biology* 80: 247–254. https://doi.org/10.1016/j.mambio.2015.03.002.

Spinage, C. A. 1973. "A review of the age determination of mammals by means of teeth, with especial reference to Africa." *East African Wildlife Journal* 11: 165–187. https://doi.org/10.1111/j.1365-2028.1973.tb00081.x.

Sumner, J., and C. R. Dickman. 1998. "Distribution and identity of species in the *Antechinus stuartii*–*A. flavipes* group (Marsupialia: Dasyuridae) in south-eastern Australia." *Australian Journal of Zoology* 46: 27–41.

van Aarde, R. J. 1985. "Age determination of cape porcupines, *Hystrix africaeaustralis*." *South African Journal of Zoology* 20: 232–236.

Van Nostrand, F. C., and A. B. Stephenson. 1964. "Age determination for beavers by tooth development." *Journal of Wildlife Management* 28 (3): 430–434.

Vaughan, T. A., J. M. Ryan, and N. J. Czaplewski. 2015. *Mammalogy.* 6th ed. Burlington, MA: Jones and Bartlett Publishers.

Venturi, F. P., C. T. Chimimba, R. J. van Aarde, and N. Fairall. 2004. "The distribution of two medically and agriculturally important cryptic rodent species, *Mastomys*

Aims

In this chapter, we discuss the different designs that can be employed to trap small mammals. We pay particular attention to the spatial arrangement of traps relative to the intended use of the data. Aside from placing traps explicitly targeting a specific animal or species (e.g., collecting museum specimens, trapping to attach tracking devices, collecting genetic samples), most small mammal traps are placed using some sort of systematic sampling regime. We introduce the importance of trapping arrangement and discuss the placement of sampling units. Then we introduce the three most common trap layouts used in small mammal studies: placing traps along transects, in grids, or in a web fashion. We also discuss advantages and disadvantages of each approach.

Arrangement of Traps

The arrangement of traps has a strong influence on what conclusions can be drawn from a study. Traps can be scattered randomly across a landscape or placed at specific locations where signs of small mammals have been detected (such as along rodent runs or at the entrance to burrows). In either case, the distance between traps will not be regular. Such irregularly placed traps may be useful under certain conditions; for example, to maximize the chances of capturing a specific species or identifying the occupant of a particular burrow. But on the whole, a haphazard arrangement will not be useful because it is inefficient, limits spatial inferences (i.e., density), and can violate the assumption of equal catchability common in many population models (Chapter 9).

The two most commonly used trapping arrangements are transects and grids. Both methods are useful, and choosing between them will depend on the types of questions being asked. A transect involves the placement of traps at regular intervals along a straight line (the "transect"). A grid, on the other hand, involves placing the traps in a number of rows and columns where the distance between adjacent traps on each row and column is identical (Fig. 7.1). A transect (Fig 7.2) can be defined by the number of traps placed along it, and the spacing between each pair of traps. A grid can be defined by the numbers of rows and columns, as well as the spacing between traps.

The critical difference between transects and grids is in the geometry of the effective area sampled by the two methods. When traps are placed in a transect, they "expose" small mammals to traps in a vastly greater area compared with the same number of traps placed in a typical grid formation (Fig. 7.2) (Pearson and Ruggiero 2003). This, of course, only holds true if traps are placed close enough to each other to allow more than one trap to fall within the home range of a small mammal (as is normally the case in ecological

Fig. 7.1. Hypothetical trapping grid, set out in a heterogeneous savanna habitat, made up of 16 traps in a 4 × 4 design. Note that this formation may result in some vegetation associations being missed by the traps. In this example, all the traps (white dots) are in open grassland.

Fig. 7.2. Hypothetical trapping transect, set out in the same location as the trapping grid (Fig. 7.1), consisting of 16 traps set out in a straight line. Note that more of the vegetation associations are covered by this design than the trapping grid (Fig. 7.1).

studies). When neighboring traps are moved farther apart, the effective area sampled by the two methods tends to converge. However, neighboring traps are not normally set hundreds of meters apart in transects or grids (at least not for rodents and other terrestrial small mammals), so this convergence is rarely experienced in the field. Instead, individual traps within transects have a greater effective area compared to traps within grids, and this has implications for choosing between the two methods.

Considering the popularity of these two methods for ecological studies of small mammal populations, surprisingly few studies have tested them against each other in the field. Nonetheless, most field studies agree that both the di-

Web Arrangement

There are two main methodologies that have been employed to enumerate populations (and hence densities) of small mammals. The first class is based on the capture-mark-recapture system that underlies the grid method outlined in the section above. In these studies, the total population is estimated using methods discussed in Chapter 9. Density is then calculated by dividing the total population by the effective sampling area, which is the trapping grid plus an additional boundary strip that is usually estimated by either movement patterns or home ranges of the animals being studied. The second class is based on distance sampling theory and trapping webs (Anderson et al. 1983). In the web-based method, traps are set along a number of traplines that radiate from a center point with traps set at regular intervals along each trapline (Fig. 7.3). Hence, the distance of any trap in the web is a fixed distance from the center point, allowing for estimates of detection functions and densities (see Chapter 9).

The trapping web is not commonly used, even though it provides better measures of density than standard grid methods (Parmenter et al. 2003) (also see Chapter 9). There have been several tests of grid-based (capture-mark-recapture) methods versus web-based (distance sampling) methods. A study by Jett and Nichols (1987) compared the densities of voles (*Microtis pennsylvanicus*) using the grid and web methods, with both giving density estimates of between 100–110 individuals per hectare. The only difference between the two methods appeared to be the greater precision of the web-based method.

Fig. 7.3. Hypothetical trapping web, set out in the same location as the trapping grid (Fig. 7.1) and the trapping transect (Fig. 7.2). Note that this design requires many more individual traps to cover the same area compared with either grids or transects.

However, since the true density of voles was unknown, it was not possible to evaluate the methods in any greater detail. A more recent study, by Parmenter et al. (2003), demonstrated that both grid and web-based methods could produce acceptable estimates of density, although grid-based methods tended to be biased at lower and higher densities.

Despite this, web-based methods are rarely used and poorly known, even by small mammal ecologists. The reasons for this are not clear but may be related to the greater trapping effort needed on trapping webs. Compared with the grid method, webs with traps in the proper locations appear to take more time to establish. This can be particularly difficult in woody environments. Additionally, it can be hard to relocate traps when they are in an irregular pattern. However, the grid method is more conducive to the estimation of various other population parameters, such as survival and recruitment, which web-based methods are not well suited for.

Conclusions and Future Directions

This chapter has summarized the different designs that can be employed to trap small mammals, particularly with respect to the spatial arrangement of traps relative to the intended use of the data. We have introduced the importance of trapping arrangement and discussed the placement of sampling units. We have discussed in depth the placement of traps along transects, in grids, and in a web fashion. We have demonstrated that transects are the best arrangement for maximizing the numbers of captures per unit of effort and for computing community composition metrics (such as species richness). In contrast, grids are necessary for the calculation of home-range movement, survival rate, recruitment, total population estimation, and relative density of different small mammal species. Web designs appear to provide better measures of population size and density of small mammals than grids, but are more difficult to implement in practice, especially in woody (i.e., non-open) vegetation. Future studies may investigate ways to make web- and distance-based designs more practicable.

REFERENCES

Anderson, D. R., K. P. Burnham, G. C. White, and D. L. Otis. 1983. "Density estimation of small-mammal populations using a trapping web and distance sampling methods." *Ecology* 64 (4): 674–680. https://doi.org/10.2307/1937188.
Avenant, N. L., and P. Cavallini. 2007. "Correlating rodent community structure with ecological integrity, Tussen-die-Riviere Nature Reserve, Free State province, South Africa." *Integrative Zoology* 2 (4): 212–219. https://doi.org/10.1111/j.1749-4877.2007.00064.x.

lungu, et al. 2011. "Impact of crop cycle on movement patterns of pest rodent species between fields and houses in Africa." *Wildlife Research* 38 (7): 603–609. https://doi.org/10.1071/WR10130.

Monadjem, A., and M. Perrin. 1998. "The effect of supplementary food on the home range of the multimammate mouse *Mastomys natalensis.*" *South African Journal of Wildlife Research* 28 (1): 1–3.

Monadjem, A., and M. Perrin. 2003. "Population fluctuations and community structure of small mammals in a Swaziland grassland over a three-year period." *African Zoology* 38 (1): 127–137.

O'Connell, M. A. 1989. "Population dynamics of Neotropical small mammals in seasonal habitats." *Journal of Mammalogy* 70 (3): 532–548.

Parmenter, R. R., T. L. Yates, D. R. Anderson, K. P. Burnham, J. L. Dunnum, A. B. Franklin, et al. 2003. "Small-mammal density estimation: A field comparison of grid-based vs. web-based density estimators." *Ecological Monographs* 73 (1): 1–26. https://doi.org/10.1890/0012-9615(2003)073[0001:smdeaf]2.0.co;2.

Patterson, B. D., P. L. Meserve, and B. K. Lang. 1990. "Quantitative habitat associations of small mammals along an elevational transect in temperate rainforests of Chile." *Journal of Mammalogy* 71: 620–633.

Pawlina, I., and G. Proulx. 1999. "Factors affecting trap efficiency: A review." In *Mammal Trapping*, edited by G. Proulx, 95–116. Sherwood Park Alberta: Alpha Wildlife Research and Management Ltd.

Pearson, D. E., and L. F. Ruggiero. 2003. "Transect versus grid trapping arrangements for sampling small-mammal communities." *Wildlife Society Bulletin* 31 (2): 454–459.

Pelikan, J. 1971. "Calculated densities of small mammals in relation to quadrat size." *Annales Zoologici Fennici* 8 (1): 3–6.

Pucek, Z., W. Jędrzejewski, B. Jędrzejewska, and M. Pucek. 1993. "Rodent population dynamics in a primeval deciduous forest (Białowieża National Park) in relation to weather, seed crop, and predation." *Acta Theriologica* 38 (2): 199–232. https://doi.org/10.4098/AT.arch.93–18.

Rancourt, S. J., M. I. Rule, and M. A. O'Connell. 2007. "Maternity roost site selection of big brown bats in ponderosa pine forests of the Channeled Scablands of northeastern Washington State, USA." *Forest Ecology and Management* 248 (3): 183–192. https://doi.org/10.1016/j.foreco.2007.05.005.

Read, K. T., K. W. J. Malafant, and K. Myers. 1988. "A comparison of grid and index-line trapping methods for small mammal surveys." *Australian Wildlife Research* 15: 673–687.

Schlesser, M., É. Le Boulengé, and N. Schtickzelle. 2002. "Can demographic parameters be estimated from capture-recapture data on small grids? A test with forest rodents." *Acta Theriologica* 47 (3): 323–332. https://doi.org/10.1007/BF03194150.

Severson, K. E. 1986. "Small mammals in modified pinyon-juniper woodlands, New Mexico." *Journal of Range Management* 39: 31–34.

Simonetti, J. A. 2006. "Microhabitat use by small mammals in central Chile." *Oikos* 56: 309–318. https://doi.org/10.2307/3565615.

Steen, H., R. A. Ims, and G. A. Sonerud. 1996. "Spatial and temporal patterns of small-rodent population dynamics at a regional scale." *Ecology* 77 (8): 2365–2372.

Stokes, M. K., N. A. Slade, and S. M. Blair. 2001. "Influences of weather and moonlight on activity patterns of small mammals: A biogeographical perspective." *Canadian Journal of Zoology* 79: 966–972. https://doi.org/10.1139/cjz-79-6-966.

Tasker, E. M., and C. R. Dickman. 2002. "A review of Elliott trapping methods for small mammals in Australia." *Australian Mammalogy* 23: 77–87. https://doi.org/10.1071/AM01077.

Tew, T. E., I. A. Todd, and D. W. MacDonald. 1994. "The effects of trap spacing on population estimation of small mammals." *Journal of Zoology, London* 233: 340–344.

Vukicevic-Radic, O., T. B. Jovanovic, R. Matic, and D. Kataranovski. 2005. "Age structure of yellow-necked mouse (*Apodemus flavicollis* Mechior, 1834) in two samples obtained from live traps and owl pellets." *Archives of Biological Sciences* 57 (1): 53–56.

Williams, B. K., J. D. Nichols, and M. J. Conroy. 2002. *Analysis and Management of Animal Populations.* Cambridge: Academic Press.

Woodman, N., R. M. Timm, N. A. Slade, and T. J. Doonan. 1996. "Comparison of traps and baits for censusing small mammals in Neotropical lowlands." *Journal of Mammalogy* 77 (1): 274–281. https://doi.org/10.2307/1382728.

8. Studying Movements

Background

Understanding a species' movements is fundamental to understanding its ecology (Turchin 1998, Hooten et al. 2017). Traditionally, studies of animal movement involved estimating the home range—the area required by animals in the normal processes of food gathering, mating, and caring for young (Burt 1943). As the ability to track animal movements improved, the complexity of home-range models increased from simple convex polygons encompassing known locations of individuals (Hayne 1949) to probabilistic utilization distributions based on sequential movements (Horne et al. 2007). Applications of these models are common in the literature, and readers are encouraged to consult reference books that rigorously cover these methods (White and Garrot 1990, Kenward 2001, Moorcroft and Lewis 2006, Hooten et al. 2017).

Studies of small mammal movements, however, are not restricted to merely estimating home-range area. Small mammal movements are studied for a myriad of reasons, such as identification of dispersal corridors (Krohne and Hoch 1999), investigation of seed caching behavior (Longland and Clements 1995), understanding effects of disturbance (Fahrig and Rytwinski 2009, Laurance et al. 2009), determination of preferred habitats (Stapp 1997, Long et al. 2013), and examination of activity patterns (Hefty and Stewart 2018). To understand how to best study this broad array of objectives, we focus this chapter on an examination of technology development and analyses with regard to studying small mammal movement in hope that this will aid researchers in selecting approaches to meet their study objectives.

As with all fields of science, in the study of animal movements there is generally a relationship between study complexity and the development of technology regarding these studies (Kays et al. 2015). Early studies of small mammal movements often focused on estimating home ranges based on spatially explicit capture-recapture data, with home-range areas being defined based on the area encompassing capture locations for individual animals (Blair 1940, Hayne 1949). Other researchers toe-clipped small mammals to uniquely identify individuals and determined subsequent locations using smoke plates (Justice 1961). Radioisotopes were also used to tag and track small mammals (Miller 1957, Kaye 1961). Other methods of tracking small mammals, i.e., powder tracking or spool-and-line tracking, were developed that enabled researchers to identify precise movement paths of small mammals but only for short periods of time (Lemen and Freeman 1985, Boonstra and Craine 1986). Use of VHF radio tags was pioneered in the 1970s (Brooks and Banks 1971, Banks et al. 1975), and miniaturization of these tags occurred rapidly thereafter, such that VHF tags are now small enough for even the smallest of small mammals (Kays et al. 2015). These VHF tags generally lack the accuracy and precision associated with powder and spool-and-line tracking, or simply following animal tracks, but they enable researchers to determine animal locations over a longer period. Today, sufficiently small GPS tags have been developed to allow precise locations to be determined for some species of small mammal, but these tags are relatively short-lived (Hefty and Stewart 2018). Continued miniaturization of GPS tags may eventually allow smaller animals to be tracked using this technology. Time of arrival approaches, often referred to as "reverse GPS," have the potential to track even the smallest of small mammal species with precision rivaling that of GPS technology but with greater transmitter lifespans (MacCurdy et al. 2009). Future technological advances offer much promise for measuring movements of small mammals.

Aims

In this chapter we introduce methods for studying movements of small mammals. For each method we provide guidelines for use, discuss advantages and disadvantages of the technique, and provide examples of past use. We then introduce some analytical procedures used for quantifying animal movement data. Finally, we offer a glimpse into potential future technologies that may further our understanding of small mammal movement.

Approaches

Most studies of animal movement can be divided into studies that measure movement using point-based methods, studies that measure continuous movement paths of animals, and studies of dispersal or colonization using molecular approaches. Although researchers can infer movement paths from point estimates, we confine our discussion to the method used to actually collect data in the field—point-based methods, path-based methods, and molecular methods.

Point-Based Methods

Many movement-based studies are centered on the idea that animal locations can be observed over time. These locations may be based on capturing the animal at a particular spatial coordinate or detecting that an animal visited a particular point. Repeated animal locations are recorded and the locations are then used to make inferences based on study objectives. These methods can be subdivided into capture-mark-recapture (CMR) methods (see Chapter 9) and electronic tracking methods, such as radio-tracking or GPS technology. Generally, CMR approaches provide fewer location estimates for individuals and typically do not have precise estimates regarding when the animal entered the trap or left its sign (e.g., uniquely identified track) at a particular location. However, the use of passive integrated transponder (PIT) tags (Smyth and Nebel 2013, Scheibler et al. 2014), date-time stamps of camera traps, or traps equipped with timing devices (Sutherland and Prevdavec 2010) can address issues regarding when an animal was in a specific location. Precise time of location is known with regard to radio-tracking or GPS-based locations.

Capture-Mark-Recapture Methods

The first studies of small mammal movements used CMR techniques in which animals were initially captured, marked, and then subsequently recaptured physically or otherwise reencountered. Each time the animal was captured or encountered, the location of the event was recorded and used for movement analyses. Although CMR studies have been used to document long-range movements of small mammals (Dickman et al. 1995), they were more generally used to measure short-range movements, home-range area (Bergstedt 1966, Leirs et al. 1996), or movements among discrete patches (Diffendorfer et al. 1995). Movement studies based on CMR techniques are often limited in scope because movements are based on the location of recapture events, which will generally be few relative to the number of locations required for most statistical models of home range.

Because animal locations are determined from recaptures or reencounters, spacing and configuration of sampling equipment, e.g., traps or tracking stations, will have direct bearing on estimates, as spatial resolution of movement cannot exceed minimum spacing between traps. Although a variety of sampling configurations have been used, uniform placement of traps in a grid pattern is perhaps the most common (see Chapter 7), and the most easily adapted, approach for deriving animal locations from capture/reencounter data. Spacing between traps is also important, with most studies placing traps at intervals ranging between 10 m and 25 m. Animal movement off of the trapping grid will not be detected using these methods; thus, grid size should be sufficient to decrease probability of marked animals venturing off of the grid. Total grid size needed and the number of traps available may provide logistical constraints on trap spacing.

Physical Recapture

Physical mark-and-recapture approaches provided the earliest estimates of small mammal movements. These techniques require that animals can be captured and uniquely marked such they can be identified when subsequently encountered. Methods for marking commonly include tattooing (Bergstedt 1966), ear-tagging (Diffendorfer et al. 1995, Andersen et al. 2000), PIT tagging (Harper and Batzli 1996), toe clipping (Monadjem and Perrin 1997), or marking pelage (Conner et al. 1999), but other methods (e.g., magnetic tags, Bergstedt 1966) may also be acceptable (see Chapter 5 for a more thorough discussion of small mammal marking methods). The basics of this approach are simple: following being marked, animals are released at the capture site, and the locations of recaptures are recorded. Locations of captures are used to determine the area traversed by the animal during the time of study. Numbers of recaptures and their spatial distribution will determine usefulness of the procedure for meeting movement objectives of the study.

A slightly different approach involves marking the animals without capture and then capturing them at a later stage. One such method is mixing Rhodamine B, a nontoxic dye, with bait and leaving it out to be consumed by rodents (Monadjem et al. 2011, Marien et al. 2018). Typical concentrations used have mixed 2 g of Rhodamine B with 1 kg of conventional bait. Once ingested, the dye becomes detectable in the whiskers of the animal for up to several weeks. At this stage, the animal needs to be captured and several whiskers removed and viewed under a fluorescent microscope with UV light at 530–585 nm. Using this method, Monadjem et al. (2011) and Marien et al. (2018) were able to show that the pest *Mastomys natalensis* readily moves between people's homes and neighboring fields and homes, with implications for the transmission of rodent-borne diseases to humans.

captured, store-onboard units may be feasible (Stevenson et al. 2013, Roeleke et al. 2016, Hefty and Stewart 2018). With animals that are short-lived or difficult to recapture, the addition of a VHF beacon to the tracking package will facilitate recovery of data if the study animal dies or the transmitter drops off of the animal. Unfortunately, the addition of a VHF beacon brings additional weight, which will reduce the number of small mammal species that can be tracked using GPS.

Researchers will do well to note that miniaturization of electronic tracking devices is largely restricted by battery technology. The smallest VHF transmitters achieve their diminutive size by using small batteries, resulting in a trade-off between longevity and range of detection. VHF tags that have the greatest battery life for a given size will either have less detection range or reduced pulse frequency, both of which result in more difficult tracking. Likewise, the smallest store-onboard GPS trackers have small batteries that only power the unit for a few days, often providing data over too brief a period to meet study demands.

Attachment of tracking devices can affect small mammal behavior. In their review on the effects of animal marking, Murray and Fuller (2000) noted that evaluation of impacts of attaching tracking devices to mammals were inconclusive, with some studies suggesting impacts on activity (Berteaux 1996) and others not (Hamley and Falls 1975, Webster and Brooks 1980). Murray and Fuller (2000) further suggested that effects on small mammal movements and behaviors may be short term.

Applications

Mikesic and Drickamer (1992) used VHF collars between 1.8 and 2.2 g to track house mice within a circular enclosure; collar weight in this case approached 10% of the animals' body weight. Derrick et al. (2010) attached 1.1 g VHF transmitter collars to cotton mice (*Peromyscus gossypinus*) and used homing to locate daytime refugia of cotton mice as part of a mesopredator exclusion study. Morris et al. (2011) used VHF radio collars that were approximately 5 g and locations derived using triangulation to estimate home ranges of cotton rats (*Sigmodon hispidus*) relative to the exclusion of mesomammal predators. Hefty and Stewart (2018) used zip ties to create a harness-type configuration to attach 5 g GPS tracking devices to quantify spatiotemporal foraging strategies in golden-mantled ground squirrels (*Callospermophilus lateralis*). GPS trackers were programed to obtain locations on 10-second intervals when the squirrels were above ground; this resulted in a lifespan for the GPS unit of only two to three days. Warren et al. (2017) implanted VHF transmitters into the peritoneal cavity of southeastern pocket gophers (*Geomys pinetis*) to estimate the gophers' home ranges. Similarly, Hansler et al. (2017) used intraperitoneal implants in maritime pocket gophers (*G. personatus maritimus*) to monitor the gophers' movements following relocation of tagged individuals.

Path-Based Methods

In this section we discuss methods allowing researchers to determine the precise path traveled by an animal. With regard to small mammals, these approaches generally refer to marking the animal in some way such that the animal leaves a readily discernable trail. Generally, path-based methods permit only short-distance tracking of individuals but may provide valuable information regarding travel corridors, climbing behavior, and foraging preferences (Lemen and Freeman 1985, Long et al. 2013).

Spool-and-line tracking was first used to track box turtles (*Terrapene carolina*); in this study, thread dispensers were attached to the turtle's carapace using tape. As the turtle moved about, the thread snagged vegetation and was pulled from the spool, leaving a trail of thread to precisely map the path taken by the turtle (Stickel 1950). The method has since be used to track a number of small mammal species. When tracking small mammals using a spool-and-line approach, researchers restrain captured animals and clip their hair to the skin at the thread attachment site (usually behind the neck and between shoulders, Boonstra and Craine 1986; or on the rump, Key and Woods 1996). A spool of thread is attached at this site using a skin-bond cement, such as the kind used for human colostomy patients. Animals are held until the glue sets and then released at the capture site. At time of release, the end of the thread should be tied firmly to an object at or near the capture site. As the animal moves about, the thread is pulled free, leaving a trail that can be readily followed (Boonstra and Craine 1986).

Tracking small mammals using fluorescent pigment (Lemen and Freeman 1985) involves carefully coating small mammals with fluorescent pigment powder (hence the name "powder tracking"), releasing the animal, and tracking the trail left behind using an ultraviolet light (Fig. 8.3). When the technique was originally introduced, the authors suggested placing captured animals in a plastic bag with the fluorescent pigment and gently shaking the bag to distribute the powder thoroughly over the small mammal's body (Lemen and Freeman 1985). As a result, this technique is often colloquially referred to as the "shake and bake" method. Because fluorescent pigments are of an exceptionally fine texture and can be readily inhaled, we do not advocate the literal shaking of small mammals in the plastic bag, even if shaking is gentle. Instead, we suggest that researchers dust the animals using a paint brush dipped in fluorescent powder, while being careful to avoid dusting the animal's head. After brushing, animals should be immediately released and

movement rates. Because of the difficulties in interpreting movement rates, we suggest those interested in measuring movement rates strive to keep time between location estimates at a minimum, as this will reduce the effect of nonlinear movements. Further, consistency in data collection protocols may facilitate comparison of relative movement rates among meaningful groups. For example, if a scientist wished to understand how forest thinning affected movement rates in a small mammal, a consistent sampling regime would only require the assumption that any differences in movement rate were due to differences in cover.

Genetic Methods

Genetic methods can be used to address a number of questions related to movement of individuals or overall patterns of gene flow. While these genetic methods do not address animal movement directly, they can provide a great amount of detail about movement indirectly. Direct methods typically will be limited in time and space due to the effort and cost of tracking or marking large numbers of individuals. Genetic approaches can overcome the space/time issue but typically at the expense of detailed individual animal movement. These inferences are considered indirect because movement is not observed directly but, rather, is inferred based on the existence of genetic structure among the individuals representing different populations (see Chapter 15 for an overview of population genetic structure). When sufficient genetic structure exists across a study area, there are a number of movement-related questions that can be addressed using genetic methods.

Early estimates of indirect dispersal were inferred through estimates of gene flow, which is based on population genetic models that relied on estimating the variance in allele frequencies among populations (e.g., F_{ST}, Wright 1951) and analogs (e.g., R_{ST}, Slatkin 1995; G_{ST}, Nei 1973) and the effective number of migrants (Nm, Wright 1951). The limitations of these approaches for quantifying movement have been addressed in detail elsewhere (see Bossart and Prowell 1998, Bohonak 1999). However, the main concerns have to do with the inability to interpret such data on ecological time frames and how well the study system fits to the specific assumptions of the genetic model under consideration (e.g., drift-migration equilibrium, Wright 1969). With improvements in the resolution of genetic markers and improved computational approaches, direct and indirect approaches have increased our ability to infer ecological-scale dispersal from molecular markers (Rojas Bonzi et al. 2019).

Assignment methods. Assignment methods use genotypic information to assign individuals to their population of origin. This is done by calculating genotypic probabilities of a specific genotype having originated in a particular population. Confidence in assignment to a population can be measured from exclusion probabilities. For each potential source population sampled, Monte Carlo simulations generate a distribution of potential genotypes based on that population's sampled allele frequencies (Cornuet et al. 1999, Paetkau et al. 2004). This generates genotype likelihood distributions that are compared to the likelihood of the target (i.e., sampled) genotype—if the empirical genotype falls within the tails of the simulated likelihoods, then it can be inferred that the empirical genotype did not originate in that population but is instead a migrant. Because these assignment methods do not assume demographic equilibrium (i.e., the balancing effect of mutation, drift, and migration on shaping allele frequencies in populations; see Chapter 15) outlier genotypes can be inferred to reflect recent migration events (Cornuet et al. 1999). Thus, the principal steps behind assignment tests are to (1) compute assignment scores one by one for genotyped individuals—assignment scores can be based on a likelihood score or a genetic distance metric, where shorter genetic distances are assumed to be the population of origin; (2) compute the exclusion score for each individual to their population of sample origin; and (3) detect immigrants from exclusion probabilities that are larger than some chosen threshold (e.g., 95% CI). A main limitation of this approach is that, strictly speaking, it assumes divergence among sampled populations (see below), but it does not estimate migration probability directly. The exclusion test of Cornuet et al. (1999) cannot effectively differentiate between a recent migrant and a genotype that is misassigned by chance (e.g., if it is a genotype that is rare in the population). Paetkau et al. (2004) corrected for this by changing the exclusion algorithm to simulate multilocus genotypes spanning 10 generations (rather than assuming loci are not linked in the genome), effectively allowing for new migrants to be distinguished from native, rare genotypes.

Assignment methods have been used to detect male-biased dispersal in agile antechinus (*Antechinus agilis*) due to males being more likely to be assigned to grids other than the one where they were captured, relative to females (Kraaijeveld-Smit et al. 2002). In contrast, female-biased dispersal was inferred from assignment tests in a larger mammal, common wombats (*Vombatus ursinus*) (Banks et al. 2002).

Assignment tests have been implemented primarily using the program GENECLASS2 (Piry et al. 2004). Sampling considerations highlight some of the methodological limitations of assignment tests like that implemented in GENECLASS2. It is assumed that population structure exists (i.e., that allele frequencies differ from sample location to sample location) and that populations can be delineated a priori. With respect to sampling, the ability to rigorously evaluate power of a genotypic dataset to discriminate among populations being compared is hampered whenever unbalanced sampling occurs, leading to individuals being assigned more often to the larger sampled populations (Wang 2016). Thus, balanced sampling is ideal. Also, data being used to generate baseline data is not independent from the samples being tested (Anderson 2010). The program assignPOP is a recent R implementation that expands assignment tests to include a machine-learning framework and can be used on genetic and nongenetic (e.g., morphometric) data (Chen et al. 2018). Another new adaptive network-based genetic assignment method implemented in R, BONE (baseline oriented network estimation) (Kuismin et al. 2020), incorporates graph inference to estimate origins and purports to identify individuals that were sampled from outside the baseline populations. While these methods have yet to be applied to small mammals, they should afford greater rigor to inferring movement using assignment methods.

Clustering methods. Another class of genetic test that can be used to infer movement are clustering algorithms that probabilistically assign individuals to genetic clusters (i.e., demes), like assignment methods, based on their individual multilocus genotypes. However, an important distinction between assignment test and clustering algorithms is the lack of any a priori assumptions about the number of genetic clusters or populations in the latter. This approach evaluates models consisting of different numbers of genetic clusters (typically denoted as K) by introducing population genetic structure using nonspatial (Pritchard et al. 2000) or spatial (e.g., C. Chen et al. 2007) Bayesian statistical methods and finding groups of individual genotypes that minimize departures from Hardy-Weinberg equilibrium and linkage disequilibrium (see Chapter 15). The popularity of this approach stems from the lack of requiring a priori knowledge of population structure. Once the number (K) of clusters are identified, the method allows for the inference of immigration (e.g., an individual's multilocus genotype being assigned to a geographically disparate cluster), as well as detection of varying levels of introgression or hybridization in individuals, which would support immigration one to two generations in the past. Illustrating this approach, Greene et al. (2017) used Bayesian clustering to identify migrants moving between anthropogenically fragmented populations of endangered beach mice (*Peromyscus polionotus trissyllepsis*).

Relatedness. Contemporary dispersal has also been estimated using patterns of relatedness between individuals sampled from different locations. Relatedness values measured from molecular markers (Chapter 15) identify not only parent-offspring or full sibling relationships (relatedness = 0.5), but they can also estimate more distant relationships categorically, such as grandparent-grandchild (0.25), half siblings (0.25), or first cousins (0.125) (Blouin 2003, Weir et al. 2006). These categorical relationships suffer from large variances creating ambiguities when differentiating among, for example, half siblings and first cousins. In contrast, kinship coefficients are measured as the probability that two alleles from the same locus sampled from two individuals are identical by descent (Blouin 2003). This ranges from zero for unrelated individuals to one for homozygous twins. The relatedness coefficient is simply two times the kinship coefficient and is most often used to measure the shared ancestry between individuals.

The use of relatedness measures holds potential power over the previously discussed assignment tests and clustering algorithms in that the latter two both require the presence of genetic structure (i.e., populations possessing different allele frequencies). However, recently diverged populations share similar genetic backgrounds making assignment tests and clustering methods unable to detect movements between them, yet genetic markers would still be powerful enough to resolved relatedness (Wang 2014). Parentage analysis has been proposed to overcome this issue (Wang 2014) where an individual source population can be determined by identifying its parents from genetic information. A major limitation here is the need for large numbers of sampled genotypes to increase the probabilities of identifying parents of individuals. Kinship has also been used to identify individuals in social groups that do not have close relationships within the group (Rollins et al. 2012). This approach has been applied to the study of larger social mammals (Archie et al. 2006) and should be useful for small mammals living in social groups. Escoda et al. (2017) applied relatedness networks to understand contemporary dispersal among neighboring river drainages in Pyrenean desman (*Galemys pyrenaicus*).

Conclusions and Future Directions

The greatest obstacle associated with quantifying movement of small mammals deals with their small size. Their small size makes them difficult to detect using passive sampling techniques, which reduces frequency of observations, which can limit inferences that can be drawn from passive sampling. When used, tracking devices such as VHF transmitters or GPS trackers must be small to avoid negatively impacting the study species. Advances in field techniques to study small mammal movements will hopefully include development of technologies leading to further miniaturization of tracking devices and their power supplies, which will ultimately permit a more thorough understanding of environmental influences on animal movements. Assignment tests, Bayesian clustering, and relatedness estimators from molecular data each hold promise for quantifying contemporary dispersal in small mammals. Like direct methods, each of these methods has limitations that will require prior understanding of the genetic background of the study system in order to determine their suitability.

REFERENCES

Andersen, D. C., K. R. Wilson, M. S. Miller, and M. Falck. 2000. "Movement patterns of riparian small mammals during predictable floodplain inundation." *Journal of Mammalogy* 81: 1087–1099.

Anderson, E. C. 2010. "Assessing the power of informative subsets of loci for population assignment: Standard methods are upwardly biased." *Molecular Ecology Resources* 10: 701–710.

Archie, E.A., C. J. Moss, and S. C. Alberts. 2006. "The ties that bind: Genetic relatedness predicts the fission and fusion of social groups in wild African elephants." *Proceedings of the Royal Society B: Biological Sciences* 273: 513–522.

Banks, E. M., R. J. Brooks, and J. Schnell. 1975. "A radio tracking study of the brown lemming (*Lemmus trimucronatus*)." *Journal of Mammalogy* 56: 888–901.

Banks, S. C., L. F. Skerratt, and A. C. Taylor. 2002. "Female dispersal and relatedness structure in common wombats (*Vombatus ursinus*)." *Journal of Zoology* 256: 389–399. https://doi.org/10.1017/S0952836902000432.

Bergstedt, B. 1966. "Home ranges and movements of the rodent species *Clethrionomys glareolus* (Schreber), *Apodemus flavicollis* (Melchior) and *Apodemus sylvaticus* (Linné) in southern Sweden." *Oikos* 17: 150–157. https://doi.org/10.2307/3564939.

Berteaux, D., F. Masseboeuf, J. M. Bonzom, J. M. Bergeron, and D. W. Thomas. 1996. "Effect of carrying a radiocollar on expenditure of energy by meadow voles." *Journal of Mammalogy* 77: 359–363.

Blair, W. F. 1940. "Home ranges and populations of the meadow vole in southern Michigan." *Journal of Wildlife Management* 4: 149–161.

Blouin, M. S. 2003. "DNA-based methods for pedigree reconstruction and kinship analysis in natural populations." *Trends in Ecology & Evolution* 18: 503–511.

Bohonak, A. J. 1999. "Dispersal, gene flow and population structure." *The Quarterly Review of Biology* 74 (1): 21–45.

Boonstra, R. and I. T. M. Craine. 1986. "Natal nest location and small mammal trapping with a spool and line technique." *Canadian Journal of Zoology* 64: 1034–1036.

Bossart, J. L., and D. P. Prowell. 1998. "Genetic estimates of population structure and gene flow: Limitations, lessons and new directions." *Trends in Ecology and Evolution* 13 (5): 202–206.

Bowers, M. A., K. Gregario, C. J. Brame, S. F. Matterm, and J. L. Dooley. 1996. "Use of space and habitats by meadow voles at the home range, patch and landscape scales." *Oecologia* 105: 107–115.

Brehme, C. S., J. A. Tracey, L. R. McClenaghan, and R. A. Fisher. 2013. "Permeability of roads to movement of scrubland lizards and small mammals." *Conservation Biology* 27: 710–720.

Briner, T., J. P. Airoldi, F. Dellsperger, S. Eggimann, and W. Nentwig. 2003. "A new system for automatic radiotracking of small mammals." *Journal of Mammalogy* 84: 571–578.

Brooks, R. J., and E. M. Banks. 1971. "Radio tracking study of lemming home range." *Communications in Behavioral Biology* 6: 1–5.

Burnett, S. E. 1992. "Effects of a rainforest road on movements of small mammals: Mechanism and implications." *Wildlife Research* 19: 95–104.

Burt, W. H. 1943. "Territoriality and home range concepts as applied to mammals." *Journal of Mammalogy* 24: 346–352.

Chen, C., E. Durand, F. Forbes, and O. François. 2007. "Bayesian clustering algorithms ascertaining spatial population structure: A new computer program and a comparison study." *Molecular Ecology Notes* 7: 747–756.

Chen, K.-Y., E. A. Marschall, M. G. Sovic, A. C. Fries, H. L. Gibbs, and S. A. Ludsin. 2018. "*assignPOP*: An R package for population assignment using genetic, nongenetic, or integrated data in a machine-learning framework." *Methods in Ecology and Evolution* 9 (2): 439–466.https://doi.org/10.1111/2041-210X.12897.

Chute, F. S., W. A. Fuller, P. R. J. Harding, and T. B. Herman. 1974. "Radio tracking of small mammals using a grid of overhead wires." *Canadian Journal of Zoology* 52: 1481–1488.

Conner, L. M., S. B. Castleberry, and A. M. Derrick. 2011. "Effects of mesopredators and prescribed fire on hispid cotton rat survival and cause-specific mortality." *Journal of Wildlife Management* 75: 938–944.

Conner, L. M., J. L. Landers, and W. K. Michener. 1999. "Fox squirrel and gray squirrel associations within minimally disturbed longleaf pine forests." *Proceedings of the Southeastern Association of Fish and Wildlife Agencies* 53: 364–374.

Conner, L. M., and B. D. Leopold. 2001. "A Euclidean distance metric to index dispersion from radiotelemetry data." *Wildlife Society Bulletin* 29: 783–786.

Cornuet, J.-M., S. Piry, G. Luikart, A. Estoup, and M. Solignac. 1999. "New methods employing multilocus genotypes to select or exclude populations as origins of individuals." *Genetics* 153: 1989–2000.

Derrick, A. M., L. M. Conner, and S. B. Castleberry. 2010. "Effects of prescribed fire and predator exclusion on refuge selection by *Peromyscus gossypinus* Le Conte (Cotton Mouse)." *Southeastern Naturalist* 9: 773–780.

Dickman, C. R., M. Oredavec, and F. J. Downey. 1995. "Long-range movements of small mammals in arid Australia: Implications for land management." *Journal of Arid Environments* 31: 441–452.

Diffendorfer, J. E., M. S. Gaines, and R. D. Holt. 1995. "Habitat fragmentation and movements of three small mammals (*Sigmodon*, *Microtus*, and *Peromyscus*)." *Ecology* 76: 827–839.

Ebensperger, L. A., R. Sobrero, V. Campos, and S. M. Giannoni. 2008. "Activity, range areas, and nesting patterns of the viscacha rat, *Octomys mimax*." *Journal of Arid Environments* 72: 1174–1183.

Escoda, L., J. Gonzales-Esteban, A. Gomez, and J. Castresana. 2017. "Using relatedness networks to infer contemporary dispersal: Applications to the endangered mammal *Galemys pyrenaicus*." *Molecular Ecology* 26: 3343–3357.

Estes-Zumpf, W. A., and J. L. Rachlow. 2009. "Natal dispersal by pygmy rabbits (*Brachylagus idahoensis*)." *Journal of Mammalogy* 90: 363–372.

Fahrig, L., and T. Rytwinski. 2009. "Effects of roads on animal abundance: An empirical review and synthesis." *Ecology and Society* 14: 21.

Fieborgm J., and L. Börger. 2012. "Could you please phrase 'home range' as a question?" *Journal of Mammalogy* 93: 890–902.

Fitzgerald, B. M., B. T. Karl, and H. Moller. 1981. "Spatial organization and ecology of a sparse population of house mice (*Mus musculus*) in a New Zealand forest." *Journal of Animal Ecology* 50: 489–518.

Goodyear, N. C. 1989. "Studying fine-scale habitat use in small mammals." *Journal of Wildlife Management* 53: 941–946.

Goosem, M. 2001. "Effects of tropical rainforest roads on small mammals: Inhibition of crossing movements." *Wildlife Research* 28: 351–364.

Goosem, M., and H. Marsh. 1997. "Fragmentation of a small-mammal community by a powerline corridor through tropical rainforest." *Wildlife Research* 24: 613–629.

Greene, D. U., J. A. Gore, and J. D. Austin. 2017. "Reintroduction of captive born beach mice: The importance of demographic and genetic monitoring." Journal of Mammalogy 98 (2): 513–522.

Hamley, J. M., and J. B. Falls. 1975. "Reduced activity in transmitter-carrying voles." *Canadian Journal of Zoology* 53: 1476–1478.

Hansler, T. P., S. E. Henke, H. L. Perotto-Baldivieso, J. A. Baskin, and C. Hilton. 2017. "Short-distance translocation as a management option for nuisance maritime pocket gophers." *Southeastern Naturalist* 16: 602–613.

Harper, S. J., and G. O. Batzli. 1996. "Monitoring use of runways by voles with passive integrated transponders." *Journal of Mammalogy* 77 (2): 364–369. https://doi.org/10.2307/1382809.

Harris, S., W. J. Cresswell, P. G. Forde, W. J. Trewhella, T. Woollard, and S. Wray. 1990. "Home-range analysis using radio-tracking data—a review of problems and techniques particularly as applied to the study of mammals." *Mammal Review* 20 (2/3): 97–123.

Hartigan, J. A. 1987. "Estimation of convex density contour in two dimensions." *Journal of the American Statistical Association* 82: 267–270.

Harvey, M. J., and R. W. Barbour. 1965. "Home range of *Microtus ochrogaster* as determined by a modified minimum area method." *Journal of Mammalogy* 46: 398–402.

Hayne, D. H. 1949. "Calculation of home range." *Journal of Mammalogy* 30: 1–18.

Hefty, K. L., and K. M. Stewart. 2018. "Novel location data reveal spatiotemporal strategies used by a central-place forager." *Journal of Mammalogy* 99: 333–340.

Hooten, M. B., D. S. Johnson, B. T. McClintock, and J. M. Morales. 2017. *Animal Movement: Statistical Models for Telemetry Data*. New York: CRC Press.

Horne, J. S., E. O. Garton, S. M. Krone, and J. S. Lewis. 2007. "Analyzing animal movements using Brownian bridges." *Ecology* 87: 1146–1152.

Jacobson, H. A., J. C. Kroll, R. W. Browning, B. H. Koerth, and M. H. Conway. 1997. "Infrared-triggered cameras for censusing white-tailed deer." *Wildlife Society Bulletin* 25: 547–556.

Justice, K. E. 1961. "A new method for measuring home ranges of small mammals." *Journal of Mammalogy* 42: 462–4701.

Kaye, S. V. 1961. "Movements of harvest mice tagged with gold-198." *Journal of Mammalogy* 42: 323–337.

Kays, R. M., C. Crofoot, W. Jetz, and M. Wikelski. 2015. "Terrestrial animal tracking as an eye on life and planet." *Science* 348 (6240): aaa2478. https://doi.org/10.1126/science.aaa2478.

Kelly, M. J. 2001. "Computer-aided photograph matching in studies using individual identification: An example from Serengeti cheetahs." *Journal of Mammalogy* 82: 440–449.

Kenward, R. E. 1987. *Wildlife Radio Tagging: Equipment, Field Techniques and Data Analysis.* London: Academic Press.

Kenward, R. E. 2001. *A Manual for Wildlife Radio Tagging.* London: Academic Press.

Key, G. E., and R. D. Woods. 1996. "Spool-and-line studies on the behavioral ecology of rats (*Rattus* spp.) in the Galápagos Islands." *Canadian Journal of Zoology* 74: 733–737.

Kie, J. G. 2013. "A rule-based ad hoc method for selecting a bandwidth in kernel home-range analyses." *Animal Biotelemetry* 1: 1–13.

Kraaijeveld-Smit, F. J. L., D. B. Linenmayer, and A. C. Taylor. 2002. "Dispersal patterns and population structure in a small marsupial, *Antechinus agilis*, from two forests analysed using microsatellite markers." *Australian Journal of Zoology* 50 (4): 325–338. https://doi.org/10.1071/ZO02010.

Krohne, D. T., and G. A. Hoch. 1999. "Demography of *Peromyscus leucopus* populations on habitat patches: The role of dispersal." *Canadian Journal of Zoology* 77: 1247–1253.

Kuismin, M., D. Saatoglu, A. K. Niskanen, H. Jensen, and M. J. Sillanpaa. 2020. "Genetic assignment of individuals to source populations using network estimation tools." *Methods in Ecology and Evolution* 11 (2): 333–344. https://doi.org/10.1111/2041-210X.13323.

Laurance, W. F., M. Goosem, and S. G. W. Laurance. 2009. "Impacts of roads and linear clearings on tropical forests." *Trends in Ecology and Evolution* 24: 659–669.

Laver, P. N., and M. J. Kelly. 2008. "A critical review of home range studies." *Journal of Wildlife Management* 72 (1): 290–298.

Leirs, H., W. Verheyen, and R. Verhagen. 1996. "Spatial patterns in *Mastomys natalensis* in Tanzania (Rodentia, Muridae)." *Mammalia* 60 (4) 545–555.

Lemen, C. A., and P. W. Freeman. 1985. "Tracking mammals with florescent pigments: A new technique." *Journal of Mammalogy* 66: 134–136.

Long, A. K., K. Bailey, D. U. Greene, C. Tye, C. Parr, H. K. Lepage, et al. 2013. "Multi-scale habitat selection of *Mus minutoides* in the Lowveld of Swaziland." *African Journal of Ecology* 51: 493–500.

Longland, W. S., and C. Clements. 1995. "Use of florescent pigments in studies of seed caching by rodents." *Journal of Mammalogy* 76: 1260–1266.

Mabee, T. J. 1998. "A weather-resistant tracking tube for small mammals." *Wildlife Society Bulletin* 26 (3): 571–574.

Mabry, K. E., and G. W. Barrett. 2002. "Effects of corridors on home range sizes and interpatch movements of three small mammal species." *Landscape Ecology* 17: 629–636.

MacCurdy, R. B., R. M. Gabrielson, E. Spaulding, A. Purgue, K. A. Cortopassi, and K. M. Fristrup. 2009. "Automatic animal tracking using matched filters and time difference of arrival." *Journal of Communications* 4: 487–495.

Marien, J., F. Kourouma, N. Magassouba, H. Leirs, and E. Fichet-Calvet. 2018. "Movement patterns of small rodents in Lassa fever-endemic villages in Guinea." *Eco-Health* 15 (2): 348–359.

Mayer, W. V. 1957. "A method for determining the activity of burrowing mammals." *Journal of Mammalogy* 38: 531.

McDonald, W. R., and C. C. St. Clair. 2004. "The effects of artificial and natural barriers on the movement of mammals in Banff National Park, Canada." *Oikos* 105: 397–407.

McGregor, R. L., D. J. Bender, and L. Fahrig. 2008. "Do small mammals avoid roads because of the traffic?" *Journal of Applied Ecology* 45 (1): 117–123. https://doi.org /10.1111/j.1365-2664.2007.01403.x.

Meserve, P. L. 1977. "Three-dimensional home ranges of cricetid rodents." *Journal of Mammalogy* 58: 549–558.

Metzgar, L. H. 1973. "Home range shape and activity in *Peromyscus leucopus*." *Journal of Mammalogy* 54: 383–390.

Mikesic, D. G., and L. C. Drickamer. 1992. "Factors affecting home-range size in house mice (*Mus musculus domesticus*) living in outdoor enclosures." *American Midland Naturalist* 127: 31–40.

Miller, L. S. 1957. "Tracing vole movements by radioactive excretory products." *Ecology* 38: 132–136.

Millspaugh, J. J., R. A. Gitzen, J. L. Belant, R. W. Kays, B. J. Keller, D. C. Kesler, et al. 2012. "Analysis of radio telemetry data." In *The Wildlife Techniques Manual*, vol. 1, 7th ed., edited by N. J. Silvy, 480–501. Baltimore: Johns Hopkins University Press.

Monadjem, A., T. A. Mahlaba, N. Dlamini, S. J. Eiseb, S. R. Belmain, L. S. Mulungu, A. W. Massawe, R. H. Makundi, K. Mohr, and P. J. Taylor. 2011. "Impact of crop cycle on movement patterns of pest rodent species between fields and houses in Africa." *Wildlife Research* 38 (7): 603–609.

Monadjem, A., and M. R. Perrin. 1997. "The effect of supplementary food on the home range of the multimammate mouse *Mastomys natalensis*." *South African Journal of Wildlife Research* 28 (1): 1–3.

Moorcroft, P. A., and M. A. Lewis. 2006. *Mechanistic Home Range Analysis*. Princeton: Princeton University Press.

Morris, G., L. M. Conner, and M. K. Oli. 2011. "Effects of mammalian predator exclusion and supplemental feeding on space use by hispid cotton rats." *Journal of Mammalogy* 92: 583–589.

Mullican, T. R. 1988. "Radio telemetry and florescent pigments: A comparison of techniques." *Journal of Wildlife Management* 52: 627–631.

Murray, D. L., and M. R. Fuller. 2000. "A critical review of the effects of marking on the biology of vertebrates." In *Research Techniques in Animal Ecology*, edited by L. Boitani and T. K. Fuller, 15–64. New York: Columbia University Press.

Myllymäki, A. 1977. "Intraspecific competition and home range dynamics in the field vole." *Oikos* 29: 553–569.

Nei, M. 1973. "Analysis of gene diversity in subdivided populations." *Proceedings of the National Academy of Sciences, USA* 70: 3321–3323.

Paetkau, D., R. Slade, M. Burden, and A. Estoup. 2004. "Genetic assignment methods for the direct, real-time estimation of migration rate: A simulation-based exploration of accuracy and power." *Molecular Ecology* 13: 55–65.

Pires, A. S., P. K. Lira, F. A. S. Fernandez, G. M. Schittini, and L. Co Oliveira. 2002.

"Frequency of movements of small mammals among Atlantic coastal forest fragments in Brazil." *Biological Conservation* 108: 229–237.

Piry, S., A. Alapetite, J.-M. Cornuet, D. Paetkau, L. Baudouin, and A. Estoup. 2004. "GENECLASS2: A software for genetic assignment and first-generation migrant detection." *Journal of Heredity* 95 (6): 536–539. https://doi.org/10.1093/jhered/esh074.

Pritchard, J., M. Stephens, and P. Donnelly. 2000. "Inference of population structure using multilocus genotype data." *Genetics* 155: 945–959.

Rémy, A., J. F. Le Galliard, G. Gundersen, H. Steen, and H. P. Andreassen. 2011. "Effects of individual condition and habitat quality on natal dispersal behaviour in a small rodent." *Journal of Animal Ecology* 80: 929–937. https://doi.org/10.1111/j.1365 -2656.2011.01849.x.

Roeleke, M., T. Blohm, S. Kramer-Schadt, Y. Yovel, and C. C. Voight. 2016. "Habitat use of bats in relation to wind turbines revealed by GPS tracking." *Scientific Reports* 6: 28961. https://doi.org/10.1038/srep28961.

Rojas Bonzi, V., C. M. Carneiro, S. M. Wisely, A. Monadjem, R. A. McCleery, B. Gumbi, and J. D. Austin. 2019. "Comparative spatial genetic structure of two rodent species in an agro-ecological landscape in southern Africa." *Mammalian Biology* 97: 64–71.

Rollins, L. A., L. E. Browning, C. E. Holleley, J. L. Savage, A. F. Russell, and S. C. Griffith. 2012. "Building genetic networks using relatedness information: A novel approach for the estimation of dispersal and characterization of group structure in social animals." *Molecular Ecology* 21: 1727–1740.

Scheibler, E., C. Roschlau, and D. Brodbeck. 2014. "Lunar and temperature effects on activity of free-living desert hamsters (*Phodopus roborovskii*, Satunin 1903)." *International Journal of Biometeorology* 58: 1769–1778.

Seaman, D. E., and R. A. Powell. 1996. "An evaluation of the accuracy of kernel density estimators for home range analysis." *Ecology* 77: 2075–2085.

Sikes, R. S., and the Animal Care and Use Committee of the American Society of Mammalogists. 2016. "Guidelines of the American Society of Mammalogists for the use of wild mammals in research and education." *Journal of Mammalogy* 97: 663–688. https://doi.org/10.1093/jmammal/gyw078.

Slatkin, M. 1995. "A measure of population subdivision based on microsatellite allele frequencies." *Genetics* 139: 457–462.

Smyth, B., and S. Nebel. 2013. "Passive integrated transponders (PIT) tags in the study of animal movement." *Nature Education Knowledge* 4: 3.

Soderquist, T. 1993. "An expanding break-away radio-collar for small mammals." *Wildlife Research* 20: 383–386.

Stapp, P. 1997. "Habitat selection by an insectivorous rodent: patterns and mechanisms across multiple scales." *Journal of Mammalogy* 78: 1128–1143.

Steinwald, M. C., B. J. Swanson, and P. M. Waser. 2006. "Effects of spool-and-line tracking on small desert mammals." *The Southwestern Naturalist* 51: 71–78.

Stevenson, C. D., M. Ferryman, O. T. Nevin, A. D. Ramsey, S. Bailey, and K. Watts. 2013. "Using GPS telemetry to validate least-cost modeling of gray squirrel (*Sciurus carolinensis*) movement with a fragmented landscape." *Ecology and Evolution* 3: 2350– 2361.

Stickel, L. F. 1950. "Populations and home range relationships of the box turtle, *Terrapene c. carolina* (Linnaeus)." *Ecological Monographs* 20: 353–378.

Stickel, L. F. 1954. "A comparison of certain methods of measuring ranges of small mammals." *Journal of Mammalogy* 35: 1–15.

Sutherland, D. R., and M. Predavec. 2010. "Universal trap timer design to examine temporal activity of wildlife." *Journal of Wildlife Management* 74: 906–909.

Swihart, R. K., and N. A. Slade. 1997. "On testing independence of animal movements." *Journal of Agricultural, Biological, and Environmental Statistics* 2: 48–63.

Thalmann, S. 2013. "Evaluation of a degradable time-release mechanism for telemetry collars." *Australian Mammalogy* 35: 241–244.

Turchin, P. 1998. *Quantitative Analysis of Movement: Measuring and Modeling Population Redistribution in Animals and Plants.* Sunderland, MA: Sinauer Associates.

Tye, C. A., D. U. Greene, W. M. Giuliano, and R. A. McCleery. 2015. "Using camera-trap photographs to identify individual fox squirrels (*Sciurus niger*) in the southeastern United States." *Wildlife Society Bulletin* 39: 645–650.

Walter, W. D., D. P. Onorato, and J. W. Fischer. 2015. "Is there a single best estimator? Selection of home range estimators using area-under-the-curve." *Movement Ecology* 3: 10. https://doi.org/10.1186/s40462-015-0039-4.

Wang, J. 2014. "Estimation of migration rates from marker-based parentage analysis." *Molecular Ecology* 23: 3191–3213.

Wang, J. 2016. "The computer program Structure for assigning individuals to populations: Easy to use but easier to misuse." *Molecular Ecology Resources* 17 (5): 981–990. https://doi.org/10.1111/1755-0998.12650.

Warren, A. E., L. M. Conner, S. B. Castleberry, and D. Markewitz. 2017. "Home range, survival, and activity patterns of the southeastern pocket gopher: Implications for translocation." *Journal of Fish and Wildlife Management* 8 (2): 544–557. https://doi.org/10.3996/032017-jfwm-023.

Webster, A. B., and R. J. Brooks. 1980." Effects of radiotransmitters on the meadow vole, *Microtus pennsylvanicus*." *Canadian Journal of Zoology* 58: 997–1001.

Wegner, J. F., and G. Merriam. 1979. "Movements by birds and small mammals between a wood and adjoining farmland habitats." *Journal of Applied Ecology* 16: 349–357.

Weir, B. S., A. D. Anderson, and A. B. Hepler. 2006. "Genetic relatedness analysis: Modern data and new challenges." *Nature Reviews Genetics* 7: 771–780.

White, G. C., and R. A. Garrott. 1990. *Analysis of Wildlife Radio-Tracking Data.* San Diego: Academic Press.

Worton, B. J. 1987. "A review of models of home range for animal movement." *Ecological Modeling* 38: 277–298.

Worton B. J. 1989. "Kernel methods for estimating the utilization distribution in home-range studies." *Ecology* 70: 164–168.

Wright, S. 1951. "The genetical structure of populations." *Annals of Eugenics* 15: 323–354.

Wright, S. 1969. *Evolution and the Genetics of Populations,* vol 2: *The Theory of Gene Frequencies.* Chicago: University of Chicago Press.

9. Estimating Population Demographics

Introduction

Fundamental to our understanding of small mammal ecology is a clear knowledge of their populations and the factors that shape those populations. Ecology has a rich history of developing methods to quantify the sizes of populations as well as the demographic rates that drive them (i.e., survival and recruitment). The data needed to quantify the population demographics of small mammals can come from a myriad of sources, including cameras (Chapter 2), radiotelemetry (Chapter 8), observations, and trapping (Chapter 7), which have been covered in other chapters. However, the selection and applicability of these methods is often a function of the goals of a project and the quantitative approaches needed to turn field data into reliable information. In many instances, people studying small mammals are interested in estimating demographic parameters. Accordingly, there is a clear need for small mammal researchers to have a basic understanding of the approaches that are available to them.

Aims

In this chapter, our aim is to provide the reader with an understanding of commonly used approaches to estimate demographics of small mammal populations. We focus on detailing the quantitative approaches for estimating population size and density, survival, and recruitment. We only present approaches that have been effective for work on small mammal populations. We discuss the assumptions, strengths, and weakness of the approaches covered in this chapter. Additionally, we provide examples of the applications of each approach from peer-reviewed literature. Due to the predominance of capture-mark-recapture data in small mammal research, most of the information presented in this chapter details a variety of ways to analyze this type of study. However, we do not belabor the mathematical underpinnings of the models and frameworks present. For a more complete understanding of estimating density and population demographics, we recommend three iconic texts: *Analysis and Management of Animal Populations* (Williams et al. 2002), *Ecological Methodology* (Krebs 2014), and the *Handbook of Capture-Recapture Analysis* (Amstrup et al. 2010).

Density

Researchers often use a statistical or target population defined by sampling (e.g., grids, transects) to infer to biological populations that encompass all individuals in an area of interest (e.g., protected area, watershed). The statistical population can be a subset of the biological population. As we discuss population estimates and demographic parameters in this chapter, we are referring to statistical populations.

Minimum Number Alive

The simplest way to estimate the density of any small mammal population is the minimum number alive estimate (MNA). This is defined as the number of individuals caught in a trapping session, including those that were not caught but must have been alive because they were caught during previous and subsequent sessions (Krebs 1966). This method does not account for the probability of undetected individuals or heterogeneity in capture probabilities (Pocock et al. 2004) and likely underestimates populations by 24% to 45% (Efford 1992). Nonetheless, MNA is a popular method of density estimation for small mammals (Efford 1992). Even recently, MNA and the number of individuals captured during a session, or minimum number of individuals (MNI), have been used for studies of rodent diseases (Carver et al. 2011, Young et al. 2014, Douglass and Vadell 2016). These "naïve density" estimators have also been useful in the study of rare and endangered species, providing conservative estimates when the rarity of detection precludes the estimation of detection parameters (McCleery et al. 2006, Harris and Macdonald 2007, La Haye et al. 2014). Pocock et al. (2004) suggested that when a simple index of density is needed, MNA should be avoided, and the MNI in a session should not be adjusted using the prior and subsequent sessions.

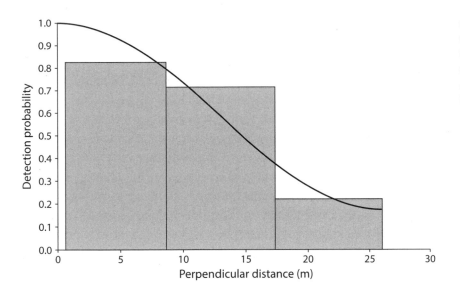

Fig. 9.1. Distance-sampling-based detection function generated from the detection of gray squirrels from line transects

Distance Sampling

Probably the most common way to estimate a wildlife population is using distance sampling (Pierce et al. 2012). This method is easily applied to larger diurnal mammal populations, which are relatively easy to observe, and is of less use for the many small mammals that are cryptic or nocturnal. Nonetheless, distance sampling has been adapted for trapping studies of small mammals that are often difficult to observe. The premise behind all distance sampling methods is that the probability of seeing an animal will decrease with distance from the observer (Burnham and Anderson 1984). The distances of observations are then used to create a histogram and to estimate the shape of a detection function, which is used to estimate the probability of detecting an animal(s) at different distances (Fig. 9.1). This detection function can then be used to estimate the number of individuals missed during sampling and to generate an estimate of the density of the population. This methodology can be used to observe animals from transects or points and has been applied to squirrels (*Sciurus* spp.) (Hein 1997, McCleery and Parker 2011), hyrax (*Dendrohyrax validus*) (Topp-Jorgensen et al. 2008), and lagomorphs (Palomares 2001), among other small mammals. While distance sampling is conceptually straightforward, estimates of precision improve with greater sample sizes, and the estimation of the detection functions can be computationally intense. Fortunately, the detection functions can be calculated using packages for the R statistical platform (see http://distancesampling.org/R/index.html) or in a popular user-friendly software package (Distance, http://distancesampling .org/). There are a number of assumptions that must be met for density estimates to be accurate: (1) sampling points and/or transects must be placed independently of animal distributions (i.e., random or systematic), (2) all objects on the line or at the point are detected, (3) animals are detected at their original location and not influenced by the observer, (4) distances are actual measurements, and (5) observations of animals are independent.

While many small mammals are not easily observed, the underpinnings of distance sampling have been adapted to estimate densities from trapping data. Anderson et al. (1983) presented the trapping web as a way to integrate the distance-based estimates of detection functions with mark-and-recapture studies. With the trapping web approach, density estimates are generated as the sum of the MNI captured within a trapping web and an estimate of the number of individuals present but not captured. Estimation of animals present but not captured is determined using a distance-based probability function. This function is estimated from the distance of captures to the center of the web (Anderson et al. 1983). The crucial step for estimating density from a trapping web (Chapter 7) is laying out the web to consist of rings of traps (>90) (Lukacs et al. 2005) that are placed with increasing distance from the center of the web (Anderson et al. 1983, Buckland et al. 2012). Similar to observing animals from a point, this design creates a gradient of capture probabilities with decreasing probability of detection as the distance from the center of the web increases. Like traditional distance sampling, the estimation of the detection function and density can be conducted using statistical packages and programs. A clear advantage to this approach to density estimation is that it addresses the issue of defining or estimating the effective trapping area. When only using the area trapped in a grid or array to estimate density, estimates tend to be positively biased because some individuals captured on the grid were likely spending time off of the grid (Dice 1938, Otis et al. 1978). The trapping web approach addresses this concern by estimating density from a point (i.e., center of the web) without assumptions of population closure (Wilson and Anderson 1985). Evaluations of trapping webs suggest that estimates of density from trapping webs perform better than grid-based estimates (Parmenter et al. 2003). This approach has been used to estimate the density of small mammals in open environments with high capture rates, such as *Perognathus* spp., *Dipodomys* spp., and *Microtus* spp. (Jett and Nichols 1987, Anderson and MacMahon 2001, Gerber and Parmenter 2015), but it has not been commonly applied in studies estimating densities of small mammals in forests. The discrepancy between the promise and use of this approach within forested environments is likely a function of the complexity associated with trap layout. In wooded systems, it takes considerable effort to lay out and find the traps in a trapping web.

Removal Method

There are a number of circumstances when researchers might want to remove small mammals from a particular area, e.g., nuisance species or population manipulation experiments. It is possible to generate population density estimates by using the number of individuals that have been captured and removed from a population. This approach is most commonly used in the study of nuisance species (i.e., *Rattus* spp., Brown et al. 1996) and rodent disease work (Di Bella et al. 2003), but it has also been applied to ecological questions (Erlinge et al. 1983). Regardless of the questions being asked, removal methods for estimating densities of small mammals follow the basic approaches outlined by Zippin (1958). The basic premise for this approach is that as more individuals are removed from a population, fewer animals will be caught. With enough effort, you can theoretically reach zero individuals caught, and the total number of individuals removed will equal the entire population. It is often difficult to remove all the small mammals from a particular area, so the population is estimated by regressing nightly removals against cumulative removals to estimate the number of individuals when the nightly removals equal zero (Erlinge et al. 1983, Brown et al. 1996). The assumptions for this approach to density estimation are (1) the population should be closed (i.e., animals are not entering or leaving the population during the study), (2) the probability of capturing an individual animal remains constant throughout the study, and (3) the probability of capture is the same for each animal (Zippin 1958). In the likely case that the probability of capturing an individual decreases over time, or if there are individuals that are trap shy in the population, the removal method will underestimate the population size (Zippin 1958).

Capture-Mark-Recapture Methods

The most commonly used, and often the most rigorous, methods for estimating the density of small mammals are capture-mark-recapture (CMR) methods. This suite of approaches can be applied to a wide breadth of study techniques. Animals can be captured with active (e.g., trapping; Chapter 3) and passive (e.g., cameras, genetic sampling; Chapter 2) methods and marked with any number of different markings (Chapter 5). The animal must then be recaptured or resighted (using the same or different methods) to determine whether the animal has been previously marked. The major assumption of these methods is that the portion of marked animals to unmarked animals in a sample is an unbiased estimate of proportion in the population (Zippin 1958, Pierce et al. 2012). There are numerous scenarios resulting in violations of this assumption. For example, if more unmarked individuals enter the population, or if marked individuals lose their markings, there would be a greater proportion of unmarked individuals in the population and estimates would be inflated. Alternatively, if marked individuals are more easily observable, or if they are more likely to go into traps after they have already been trapped (trap happy), the sample will be bias toward marked individuals and density estimates will be biased downward. For a more thorough examination of these assumptions and CMR methods in general, see the texts by Williams et al. (2002) and Pierce et al. (2012).

Models that use CMR data to can be broadly categorized into open or closed models. Closed models assume the population size does not change (i.e., no births, deaths, immigration, or emigration) during the period sampled. Closed models only estimate population size or density. Opened models allow the size of the population to change over the study and can estimate demographic parameters other than abundance or density.

Closed Models

The oldest and most commonly used closed model to estimate density of small mammals from two samples, preferably during a short time frame, is the Lincoln-Petersen estimator. During the first sample period (i.e., trapping), small mammals are marked, but they do not have to be individually marked. During the second sampling period, a random sample is obtained (recaptured or resighted such that all individuals have the same probability of being captured) and the number of marked and unmarked individuals are tallied. Because the assumption can be exceedingly difficult to meet, many of the advances in CMR methods address the assumption of equal catchability. The population is then estimated by taking the number of individuals marked in the first sample multiplied by the total number of individuals captured in the second, divided by the number of marked individuals in the second sample. While this approach is intuitive, it has been shown to overestimate the size of small populations (Krebs 2014). To reduce this bias, Seber (1982) and others have suggested modifications to this original formulation. Additionally, it can be difficult to meet the assumption of equal trappability for different sexes and sizes of individuals from the same population. Nonetheless, ecologists still use the original and modified formulation of the Lincoln-Petersen approach to estimate density of small mammals. Recent examples include estimates of small mammal abundance in shrublands of southern California (Diffendorfer et al. 2012), squirrels in Alaska (Smith 2012), and murid rodents in East Africa (Borremans et al. 2011).

Small mammals are often trapped for more than two sampling occasions. With repeated sampling, it is possible to extend the Lincoln-Petersen method to include a series of recapture data. Density from this series of recapture sessions can be individually calculated as Lincoln-Petersen estimates and averaged. More commonly, the data are given a weighted average, as first proposed by Schnabel (1938). A number of adjustments have been made to the Schnabel formulation (Schnabel 1938). For example, Schumacher and Eschmeyer (1943) suggested estimating density using a regression of the number of individuals previously marked by the portion of marked individuals in each sample. This approach appears to be more robust than Schnabel's original formulation (Krebs 2014), but it and other variants still have many of the same assumptions and limitations of the Lincoln-Peterson method. The advantage of the Schnabel method is that it is easier to determine whether the assumptions of catchability are violated because a plot of the portion of marked animals during a session against the number of animals previously marked should be linear (Krebs 2014). Like the Lincoln-Peterson method, the Schnabel method and its variants are still commonly used to estimate the density of small mammals (Borchert and Borchert 2013, Young et al. 2014, Degrassi 2018).

The Schnabel method can be used to detect unequal trappability of small mammals, but it does not provide researchers a way to address the issue if it occurs. This issue was brought to the forefront by Otis et al. (1978), who described several multiple-capture models that accounted for capture heterogeneity (unequal capture rates). These models captured three potential sources of variability in capture rates: (1) time effects (e.g., variation due to weather or environment), (2) behavioral effects (e.g., trap happy, trap shy), and (3) individual variation (e.g., sex, age). From these potential sources of variation, Otis et al. (1978) and White et al. (1978) then developed eight potential models of capture heterogeneity in Program CAPTURE. For a full review of models in Program CAPTURE, see the authoritative chapter by Chao and Huggins (2005). The program also allows researchers to evaluate each model's ability to explain capture heterogeneity based on a measure of parsimony (Akaike's Information Criteria, AIC) (Rexstad and Burnham 1991). Program CAPTURE also integrates jackknife estimators that consider subsets of data by deleting capture occasions (Chao and Huggins 2005). This approach has only limited theoretical support but performs well in simulations (Chao and Huggins 2005). Program CAPTURE, which has now been integrated into Program MARK (White and Burnham 1999), has been the tool of choice and a workhorse for many ecologists estimating density of closed populations of small mammals (see Menkens and Anderson 1988, Nupp and Swihart 2000, Hammond and Anthony 2006, Ruiz-Capillas et al. 2013). While Program CAPTURE made big strides in addressing capture heterogeneity, it has been criticized based on the performance of its density estimates (Menkens and

Anderson 1988), the parametrization of some of the candidate models, and the inability to include covariates to model capture heterogeneity (Chao and Huggins 2005). Due to these short comings, Manly et al. (2005) recommend the Huggins approach for closed models because of its ability to include covariates and its flexibility and accessibility using Program MARK or RMark (Laake and Rexstad n.d.). The Huggins approach has also been applied to estimate the density of all different forms of trappable small mammals, from soricids to lagomorphs. Specific examples include estimates of the populations of endangered species (Schmidt et al. 2011) and the response of small mammals to fire (Converse et al. 2006), mesopredators (Eagan et al. 2011), and agricultural edges (Hurst et al. 2014).

Spatial Consideration for Capture Grids

An inherent problem for grid-based density estimates is that it can be difficult to know the effective area that has been trapped by the grid because animals are likely to move on and off of the grids. There are several ways to address this problem. The simplest way is to add a boundary strip around the grid that is based on the movements of the animals being studied, usually half of the radius of the animal's home range during the trapping occasions. This estimate can be derived from animal movements between traps or from radiotagging. A common way to estimate a boundary strip that appears to perform well in field research and simulation studies (Tanaka 1972, Wilson and Anderson 1985, Keiter et al. 2017) is to use the mean maximum distance moved (MMDM). This is calculated by averaging the maximum distances of each animal captured more than once (Krebs 2014). This technique has recently been applied to studies on fox squirrels (*Sciurus niger*) (Greene and McCleery 2017), Keen's mice (*Peromyscus keeni*) (Eckrich et al. 2018), and southern redbacked voles (*Myodes gapperi*) (Sullivan and Sullivan 2017), among others. A second approach developed on small mammals, but that is not commonly used, relies on nested grids (Otis et al. 1978). Nested grids divide large grids into a suite of smaller "nested" grids. This approach requires large grids (e.g., 16 × 16) and assumes a reduction in bias occurs with increasing size of the grid and boundary strip. The change in density estimates of the nested grids can be used to estimate the boundary strip for the entire grid (Krebs 2014). Density estimates using the nested grid approach can be calculated using Program CAPTURE. The nested grid approach has been used to estimate the density of endangered woodrats (Humphrey 1988) and desert rodent communities (Parmenter et al. 2003); however, it takes considerable effort to maintain the large grids needed for the nested approach, which may explain why it is not commonly used in small mammal research (Krebs et al. 2011).

Spatially Explicit Capture-Recapture Methods

Similar to web-based grids, the increasingly popular spatially explicit capture-recapture (SECR) methods use capture locations of individuals to model animal movements on rectangular (or other shapes) grids and to generate detection probabilities. Rigorous estimates of density are generated by using movement information to estimate the probability that an animal will be captured in a trap on the grid and to calculate an effective sampling area (Efford and Fewster 2013). These models require that at least some individuals are captured at multiple locations. The advantages of SECR methods are that they allow for variation in individual exposure to traps and define the geographic area of capture. Population density estimates using this approach appear to be relatively unbiased (Efford and Fewster 2013). However, SECR methods need a great number of small mammal capture events to provide reasonable estimates and are limited by single-catch trapping devices (e.g., Sherman traps) that might limit detection. Precise density estimates of small mammals are likely to require resurveys, fine tuning of trap spacing, and traps that can record multiple captures (e.g., cameras, nest boxes) (Romairone et al. 2018). For these reasons, SECR methods are rarely applied to the study of small mammals; however, Romairone et al. (2018) used this method with common voles (*Microtus arvalis*) to provide an example of how the method can be applied to small mammals. There are a number of packages on the R platform (e.g., unmarked, SECR, oSCR, SPACECAP R) that can be used to generate SECR estimates.

Open Models

Open models, which allow the size of a population to change over time, can be used to estimate several demographic parameters for small mammal populations. Some of the most widely used open models can estimate parameters such as apparent survival and recruitment. We use the term "apparent survival" because in most models it is not possible to distinguish between emigration and mortality. Recruitment can be defined as the number of individuals entering the population.

To collect data for standard open population models, animals need to be captured, individually marked (Chapter 5), and released during at least three sampling periods. During each sampling period, there are usually multiple capture occasions (e.g., four nights of consecutive trapping). The length of the sampling period should be short relative to time between sampling periods (e.g., three months). While it is possible to apply open models to data with unequal intervals between sampling occasions, we recommend keeping consistent time intervals between sampling periods.

The original model for open populations was the Jolly-Seber (JS) model (Jolly 1965, Seber 1965). The JS model estimates population size, capture probability, apparent survival, and recruitment. The population size is estimated based on the relationship of marked and unmarked individuals in the population (Pierce et al. 2012). The JS model has been criticized because it assumes the same probability for the capture of marked and unmarked individuals, which can lead to bias in population size estimates (Pledger et al. 2010). The Cormack-Jolly-Seber (CJS) model addressed this issue by only accounting for information at the first capture of the individual, while the JS method uses the information (0s) before the first capture (Kéry and Schaub 2012). Accordingly, the more restricted CJS model can only estimate apparent survival and recapture probabilities and does not estimate the population size. The CJS model does not assume that unmarked individuals in the population have the same probability of capture as marked (Schwarz and Arnason 2006), but many of the other assumptions of this model are similar to the JS model. Both models assume (1) marks are not lost and are recorded without error, (2) survival is the same among marked and unmarked individuals, (3) the study area is constant, (4) sampling is done instantaneous (i.e. over a short period of time) and captured animals are released immediately, and (5) captured and recaptured animals represent a random sample from the same population (Pierce et al. 2012, Kéry and Schaub 2012). Both models have been modified form their original form to provide more flexibility and robust estimations. For example, the JS model has been modified to allow for heterogeneity in capture and survival rates (Pledger et al. 2003), small sample sizes, and losses (death) during capture, and to add a population growth rate parameter (Schwarz and Arnason 2006, Krebs 2014). The CJS model has been adapted to include covariates of time, groups (sex, age), and individual variation (Nichols 2005). The original models and their derivations are still used heavily for estimating small mammal demographics. For recent examples of the JS model, see the work on red-backed voles (*Myodes gapperi*) by Sullivan et al. (2017) and see Ostfeld et al.'s (2018) examination of small mammal population cycles and disease prevalence. Recent examples of the use of CJS models include Wangdwei et al.'s (2013) work on plateau pikas (*Ochotona curzoniae*) and Kovacs et al.'s (2012) investigation of the response of bush rats (*Rattus fuscipes*) to invasive predators. The computation of the JS and CJS models is commonly performed in Program MARK (White and Burnham 1999) or packages for the R platform, such as RMark (Laake and Rexstad n.d.) and marked (Laake et al. 2013).

One particularly useful modification of the JS model for estimating small mammal demographics has been Pradell's temporal symmetry approach. In addition to the traditional forward time modeling of populations, where earlier trapping sessions inform later ones, this approach also utilizes reverse time modeling, where capture histories are considered in reserve to estimate recruitment by determining when animals entered the population (Williams et al. 2002). Pradell's temporal symmetry approach also allows for different model parameterizations to estimate apparent survival, recruitment, capture probabilities, and population growth rates (Williams et al. 2002). Furthermore, this approach allows for inclusion of covariates to model the influence of factors such as sex and age on demographic estimates. Some examples of the application of Pradell's temporal symmetry to small mammal populations include determining factors that were driving the decline of endangered woodrats (*Neotoma floridana smalli*) (McCleery et al. 2013), the impacts of fire on at-risk small mammals (Griffiths et al. 2015), and the influence of small mammal density dependence on recruitment (Pinot et al. 2016).

Robust Design

Closed population estimation, generated over a short timeframe, can produce rigorous estimates of population abundance during a snapshot in time. In contrast, we have presented several open population models that estimate changes in demographic parameters over time. One of the shortcomings of these open models this they do not allow individual variation in capture probabilities, which can bias abundance estimates (Williams et al. 2002). The robust design addresses this problem by integrating open and closed CMR approaches into a single design where populations are sampled over short, closed periods separated by longer open periods. By providing additional information during the short-term sampling, this approach allows for a more robust estimation of the demographic parameters estimated by open models (Williams et al 2002). Specifically, this "robust design" (Pollock 1982) uses closed population models like those detailed by Otis et al. (1978) to estimate abundance and then pools the data (animals were captured or not), uses a CJS model to estimate survival, and uses estimates of survival and abundance to estimate recruitment (Nichols 2005). For the closed models in the robust design, we recommend the use of the Huggins models (Kendall 2001) because of their flexibility. The robust design has been used extensively in long-term studies of rodent populations; however, rare or elusive small mammals often do not yield the sample sizes needed to implement this approach. Some recent applications of the robust design to small mammal populations include examinations of fox squirrel (*Sciurus niger*) (Greene and McCleery 2017) and cyclic vole (*Myodes glareolus*) (Johnsen et al. 2017) populations, understanding the influence of management practices on small mammals (King et al. 2014, Gasperini et al. 2016, Wayne et al. 2016), and understanding the influence of predators on brown lemmings (*Lemmus trimucronatus*) (Fauteux et al. 2016). All the authors of these studies used Program MARK (White and Burnham 1999) or the RMark package (Laake and Rexstad n.d.) in R, which is based on Program MARK.

True Survival

Determining survival rates and causes of mortality is critical to an understanding of what drives small mammal populations. However, most models that rely on mark-recapture data can only estimate apparent survival because they cannot distinguish two inherently different processes: mortality and immigration. To separate these processes, small mammal researchers often turn to radiotelemetry (Chapter 8). The mortality events identified from radiotelemetry or another tracking method can then be turned into estimates of survival. Possibly the simplest approach to estimating survival is to generate a constant, or finite, estimate of survival over time, where survival equals the number of animals at the end of the study period divided by the number of animals at the beginning of the study period. This basic approach has been advanced to account for the number of telemetry days animals were at risk of mortality and to allow for survival to vary among time intervals (e.g., weeks, months) with an assumed constant daily rate (Mayfield 1975, Heisey and Fuller 1985). Additionally, it is possible to partition survival for different causes of death or attribute classes (e.g., sex, age). These analyses can be performed with a number of computer programs that have been incorporated in Program MARK (White and Burnham 1999). Small mammal researchers have used this approach to compare the survival of urban and rural fox squirrels (McCleery et al. 2008) and cotton rats exposed to mesopredators and fire (Conner et al. 2011), and to study environmental factors influencing the survival rates of muskrats (*Ondatra zibethicus*) (Ahlers et al. 2010). Some potential constraints of this approach are that it assumes constant rates of survival among individuals of the same group and constant survival within selected time intervals (Williams et al. 2002).

A potentially more flexible approach is to estimate survival rates based on an individual's time of death and the amount of time that it had an active radio collar (Krebs 2014). This approach produces survival curves that describe the shape of a survival function over time based on the radio-tagged animals in the study (Skalski et al. 2005). Survivorship curves can be analyzed using parametric and nonparametric approaches. Parametric approaches assume a smooth function can be fit to survivorship data but, because of the unpredict-

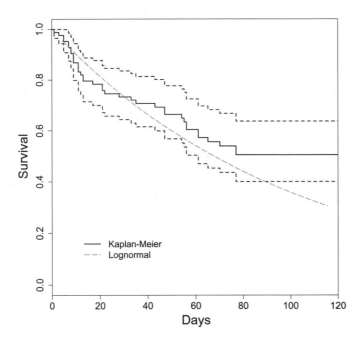

Fig. 9.2. Parametric (lognormal) and nonparametric (Kaplan-Meier) survival curves with 95% CI (*dotted black lines*) for woodrats over 120 days

ability of mortality events and often small sample sizes (Williams et al. 2002) associated with data from small mammals, may not provide a good fit. An alternative nonparametric approach to survival curves is the Kaplan-Meier method (Fig. 9.2) for staggered entry data (Pollock et al. 1989). This method provides an estimation of survival rates from staggered entry data over time, which can be expressed graphically and used to compare different groups or treatments (Williams et al. 2002). The approach has been used to compare survival of wild captivity-released woodrats (*Neotoma* spp.) (McCleery et al. 2013, Blythe et al. 2015), as well as to estimate survival of white-footed mice (*Peromyscus leucopus*) (Collins and Kays 2014) and southeastern pocket gophers (*Geomys pinetis*) (Warren et al. 2017).

Reproductive Parameters

Like survival rate estimation, our knowledge of reproductive parameters from capture-recapture studies is often limited because, in most cases, it is not possible to determine if individuals recruited into the population immigrated or were born to individuals within the trapping area. This problem is exacerbated with small mammals because they often have high rates of mortality between birth and weaning and they do not enter capture-recapture studies until they are old enough to venture out of natal nests. The estimation of reproductive parameters is relatively straightforward (see Skalski et al. 2005) but logistically challenging. One basic reproductive measure of population is productivity, or the ratio of young to adults; this can also be expressed as the number of juvenile females to adult females. Deriving this ratio requires some information about the number births and the sex ratios of young. Earlier in the book, we presented an invasive method (animal sacrificed) to calculate litter sizes using placental scars and embryo counts (Chapter 6). To get this information without sacrificing individual small mammals, ecologists have often used controlled environments such as labs and captive-breeding programs (Lloyd and Christian 1969, Daly et al. 1984, Buffenstein 2005); however, these measures may not be representative of wild populations and conditions. To get more realistic measures of reproductive parameters, ecologists have estimated litter sizes of small mammals (e.g., Microtinae) from live births in traps (Innes 1978). An alternative approach for collecting these reproductive parameters (fecundity, sex ratios, juvenile survival) of small mammals is the use of artificial nests, often in the form of nest boxes (Goundie and Vessey 1986, Rose and Walke 1988, Rose 2008) (see Fig. 2.3). Ideally, small mammals will take up residence in the artificial nests, and their reproductive output and survival of their young can be monitored through routine checks. Artificial nests can be monitored by looking through windows or trap doors or by using cameras. Increasingly, cameras appear to be the least disruptive way to monitor the reproductive outputs of small mammals from artificial nests (Forsman et al. 2009, McCleery 2009, Sotola and Garneau 2014).

Genetic Estimates of Population Size

Many small mammals are highly secretive or elusive, making it difficult to obtain accurate estimates of population size using traditional methods of marking and recapturing (Luikart et al. 2010). Genetic methods that rely on noninvasively sampled hairs, feces, or urine (Chapter 2) may help to resolve some of these challenges of working on small mammals and may reduce the impact of handling on subsequent catchability (Petit and Valière 2006). Genetic methods require "genetic fingerprints," typically consisting of multilocus genotypes using high-resolution genetic markers, such as microsatellites or single nucleotide polymorphisms (SNPs), that can distinguish individuals (Chapter 15). Genotypes can be collected over multiple temporal recapture events and treated as typical permanent tags, as discussed above, to estimate demographic parameters such as population size and survival. Examples include studies on spotted-tailed quoll (*Dasyurus maculatus*) (Ruibal et al. 2009) and Cabrera vole (*Microtus cabrerae*) (Proença-Ferreira et al. 2019).

A potential benefit of genetic approaches is that longitudinal sampling is not necessary, as genotypes matched across contemporaneously samples can be treated as recaptures, and this has the advantage of better approximating the population closure assumption (Miller et al. 2005, Luikart et al. 2010). Capwire (Pennell et al. 2013) is a maximum likelihood estimator developed for "single sample" studies that provides 95% confidence intervals on estimates and performs log-likelihood ratio tests for model selection (i.e., equal-capture vs. two-innate rates) and appears to work well on small populations (e.g., N_C <100) and when recapture rates are heterogeneous.

Rarefaction analyses have also been used from genetic tag data to estimate abundance in an area by plotting the cumulative number of newly recorded individuals against the total number sampled (Frantz and Roper 2006). However, this rarefaction method has not been frequently used on small mammals.

Census methods have also been based on pedigree reconstructions from genetic markers. The Creel-Rosenblatt estimator (CRE) (Creel and Rosenblatt 2013) incorporates the sum of unique sampled individuals (N_s), number of breeders (B_s), number of individuals inferred from pedigree reconstruction (N_{in}), and the estimated number of individuals that did not breed nor were sampled (rendering them invisible to pedigree reconstruction) into the population estimate. This approach has been used to estimate census size of larger populations (Spitzer et al. 2016); however, a modified version of it was used to estimate the size of the breeding female population in mountain pygmy possum (*Burramys parvus*) (Hettiarachchige and Huggins 2018).

The use of genetic samples in mark-recapture studies remains underutilized in studies on small mammals (Jung et al. 2020) yet holds promise for not only estimating N but also for providing insight into individual survival, emigration rates, and finite rates of population change. While beyond the scope of this chapter, a good overview of the issues and applications of noninvasive genetic samples in capture-recapture studies can be found in Lukacs and Burnham (2005).

Conclusions and Future Direction

Understanding the demographic dynamics of small mammals has been foundational in our understanding of their ecology. Small mammal ecologists have benefited from, and been instrumental in, the development of analytical methods that are now widely used in the estimation of the demographic parameters of animal populations. In the last several decades, there has been a steady stream of advances in the estimates of population parameters that reduced the probability of violated assumption and improved accuracy. Some of these advances work well for small mammal populations (i.e., Pradell's temporal symmetry) while other approaches, such as SECR methods, clearly do not. While we expect to continue to see advances in the analysis of animal populations, we suspect that many of these advances will only be applicable to common small mammal species that provide the amount of data needed to fit parameters to these models. Moving forward, we expect small mammal researchers to rely more on genetics and using the models presented in this chapter within a Bayesian framework (see Mamba et al. 2019). Due to their flexibility and ability to estimate parameters with previous, limited, and extensive datasets, we expect to see a broader use of Bayesian approaches to estimate small mammal population parameters in the years to come.

REFERENCES

Ahlers, A. A., R. L. Schooley, E. J. Heske, and M. A. Mitchell. 2010. "Effects of flooding and riparian buffers on survival of muskrats (*Ondatra zibethicus*) across a flashiness gradient." *Canadian Journal of Zoology* 88 (10): 1011–1020. https://doi.org/10.1139/z10-069.

Amstrup, S. C., T. L. McDonald, and B. F. J. Manly. 2010. *Handbook of Capture-Recapture Analysis*. Princeton: Princeton University Press.

Anderson, C. J., and J. A. MacMahon. 2001. "Granivores, exclosures, and seed banks: Harvester ants and rodents in sagebrush-steppe." *Journal of Arid Environments* 49 (2): 343–355. https://doi.org/10.1006/jare.2000.0781.

Anderson, D. R., K. P. Burnham, G. C. White, and D. L. Otis. 1983. "Density estimation of small-mammal populations using a trapping web and distance sampling methods." *Ecology* 64 (4): 674–680. https://doi.org/10.2307/1937188.

Blythe, R. M., T. J. Smyser, S. A. Johnson, and R. K. Swihart. 2015. "Post-release survival of captive-reared Allegheny woodrats." *Animal Conservation* 18 (2): 186–195. https://doi.org/10.1111/acv.12158.

Borchert, M., and S. M. Borchert. 2013. "Small mammal use of the burn perimeter following a chaparral wildfire in southern California." *Bulletin Southern California Academy of Sciences* 112 (2): 63–73.

Borremans, B., H. Leirs, S. Gryseels, S. Guenther, R. Makundi, and J. G. de Bellocq. 2011. "Presence of Mopeia virus, an African arenavirus, related to biotope and individual rodent host characteristics: Implications for virus transmission." *Vector-Borne and Zoonotic Diseases* 11 (8): 1125–1131. https://doi.org/10.1089/vbz.2010.0010.

Brown, K. P., H. Moller, J. Innes, and N. Alterio. 1996. "Calibration of tunnel tracking rates to estimate relative abundance of ship rats (*Rattus rattus*) and mice (*Mus musculus*) in a New Zealand forest." *New Zealand Journal of Ecology* 20 (2): 271–275.

Buckland, S. T., D. R. Anderson, K. P. Burnham, and J. L. Laake. 2012. *Distance Sampling: Estimating Abundance of Biological Populations*. Berlin: Springer Science & Business Media.

Buffenstein, R. 2005. "The naked mole-rat: A new long-living model for human aging research." *Journals of Gerontology Series a-Biological Sciences and Medical Sciences* 60 (11): 1369–1377. https://doi.org/10.1093/gerona/60.11.1369.

Burnham, K. P., and D. R. Anderson. 1984. "The need for distance data in transect counts." *Journal of Wildlife Management* 48 (4): 1248–1254. https://doi.org/10.2307/3801785.

Carver, S., J. T. Trueax, R. Douglass, and A. Kuenzi. 2011. "Delayed density-dependent prevalence of *Sin Nombre virus* infection in deer mice (*Peromyscus maniculatus*) in central and western Montana." *Journal of Wildlife Diseases* 47 (1): 56–63. https://doi.org/10.7589/0090-3558-47.1.56.

Chao, A., and Huggins R. (2005). "Modern closed-population capture-recapture models." In *Handbook of Capture-Recapture Analysis*, edited by S. C. Amstrup, T. L. McDonald, and B. F. J. Manly, 58–87. Princeton: Princeton University Press.

Collins, C. R., and R. W. Kays. 2014. "Patterns of mortality in a wild population of white-footed mice." *Northeastern Naturalist* 21 (2): 323–336. https://doi.org/10.1656/045.021.0213.

Conner, L. M., S. B. Castleberry, and A. M. Derrick. 2011. "Effects of mesopredators and prescribed fire on hispid cotton rat survival and cause-specific mortality." *Journal of Wildlife Management* 75 (4): 938–944. https://doi.org/10.1002/jwmg.110.

Converse, S. J., W. M. Block, and G. C. White. 2006. "Small mammal population and habitat responses to forest thinning and prescribed fire." *Forest Ecology and Management* 228 (1–3): 263–273. https://doi.org/10.1016/j.foreco.2006.03.006.

Creel, S., and E. Rosenblatt. 2013. "Using pedigree reconstruction to estimate population size: Genotypes are more than individually unique marks." *Ecology and Evolution* 3: 1294–1304.

Daly, M., M. I. Wilson, and P. Behrends. 1984. "Breeding of captive kangaroo rats, *Dipodomys merriami* and *D. microps*." *Journal of Mammalogy* 65 (2): 338–341. https://doi.org/10.2307/1381177.

Degrassi, A. L. 2018. "Hemlock woolly adelgid invasion affects microhabitat characteristics and small mammal communities." *Biological Invasions* 20 (8): 2173–2186. https://doi.org/10.1007/s10530-018-1694-3.

Di Bella, C., F. Vitale, G. Russo, A. Greco, C. Milazzo, G. Aloise, et al. 2003. "Are rodents a potential reservoir for *Leishmania infantum* in Italy?" *Ibex Journal of Mountain Studies* 7: 125–129.

Dice, L. R. 1938. "Some census methods for mammals." *Journal of Wildlife Management* 2 (3): 119–130. https://doi.org/10.2307/3796432.

Diffendorfer, J., G. M. Fleming, S. Tremor, W. Spencer, and J. L. Beyers. 2012. "The role of fire severity, distance from fire perimeter and vegetation on post-fire recovery of small-mammal communities in chaparral." *International Journal of Wildland Fire* 21 (4): 436–448. https://doi.org/10.1071/wf10060.

Douglass, R. J., and M. V. Vadell. 2016. "How much effort is required to accurately describe the complex ecology of a rodent-borne viral disease?" *Ecosphere* 7 (6). https://doi.org/10.1002/ecs2.1368.

Eagan, T. S., II, J. C. Beasley, Z. H. Olson, and O. E. Rhodes. 2011. "Impacts of generalist mesopredators on the demography of small-mammal populations in fragmented landscapes." *Canadian Journal of Zoology* 89 (8): 724–731. https://doi.org/10.1139/z11-045.

Eckrich, C. A., E. A. Flaherty, and M. Ben-David. 2018. "Functional and numerical responses of shrews to competition vary with mouse density." *PLoS ONE* 13 (1). https://doi.org/10.1371/journal.pone.0189471.

Efford, M. 1992. "Comment: Revised estimates of the bias in the minimum number alive estimator." *Canadian Journal of Zoology* 70 (3): 628–631. https://doi.org/10.1139/z92-093.

Efford, M. G., and R. M. Fewster. 2013. "Estimating population size by spatially explicit capture-recapture." *Oikos* 122 (6): 918–928. https://doi.org/10.1111/j.1600-0706.2012.20440.x.

Erlinge, S., G. Goransson, L. Hansson, G. Hogstedt, O. Liberg, I. N. Nilsson, et al. 1983. "Predation as a regulating factor on small rodent populations in southern Sweden." *Oikos* 40 (1): 36–52. https://doi.org/10.2307/3544197.

Fauteux, D., G. Gauthier, and D. Berteaux. 2016. "Top-down limitation of lemmings revealed by experimental reduction of predators." *Ecology* 97 (11): 3231–3241. https://doi.org/10.1002/ecy.1570.

Forsman, E. D., J. K. Swingle, and N. R. Hatch. 2009. "Behavior of red tree voles (*Arborimus longicaudus*) based on continuous video monitoring of nests." *Northwest Science* 83 (3): 262–272. https://doi.org/10.3955/046.083.0309.

Frantz, A. C., and T. J. Roper. 2006. "Simulations to assess the performance of different rarefaction methods in estimating population size using small datasets." *Conservation Genetics* 7: 315–318. https://doi.org/10.1007/s10592-006-9125-x.

Gasperini, S., A. Mortelliti, P. Bartolommei, A. Bonacchi, E. Manzo, and R. Cozzolino. 2016. "Effects of forest management on density and survival in three forest rodent species." *Forest Ecology and Management* 382: 151–160. https://doi.org/10.1016/j.foreco.2016.10.014.

Gerber, B. D., and R. R. Parmenter. 2015. "Spatial capture-recapture model performance with known small-mammal densities." *Ecological Applications* 25 (3): 695–705. https://doi.org/10.1890/14-0960.1.

Goundie, T. R., and S. H. Vessey. 1986. "Survival and dispersal of young white-footed mice born in nest boxes." *Journal of Mammalogy* 67 (1): 53–60. https://doi.org/10.2307/1381001.

Greene, D. U., and R. A. McCleery. 2017. "Reevaluating fox squirrel (*Sciurus niger*) population declines in the southeastern United States." *Journal of Mammalogy* 98 (2): 502–512. https://doi.org/10.1093/jmammal/gyw186.

Griffiths, A. D., S. T. Garnett, and B. W. Brook. 2015. "Fire frequency matters more than fire size: Testing the pyrodiversity-biodiversity paradigm for at-risk small mammals in an Australian tropical savanna." *Biological Conservation* 186: 337–346. https://doi.org/10.1016/j.biocon.2015.03.021.

Hammond, E. L., and R. G. Anthony. 2006. "Mark-recapture estimates of population parameters for selected species of small mammals." *Journal of Mammalogy* 87 (3): 618–627. https://doi.org/10.1644/05-mamm-a-369r1.1.

Harris, D. B., and D. W. Macdonald. 2007. "Population ecology of the endemic rodent *Nesoryzomys swarthi* in the tropical desert of the Galápagos Islands." *Journal of Mammalogy* 88: 208–219.

Hein, E. W. 1997. "Demonstration of line transect methodologies to estimate urban gray squirrel density." *Environmental Management* 21 (6): 943–947. https://doi.org/10.1007/s002679900078.

Heisey, D. M., and T. K. Fuller. 1985. "Evaluation of survival and cause-specific mortality-rates using telemetry data." *Journal of Wildlife Management* 49 (3): 668–674. https://doi.org/10.2307/3801692.

Hettiarachchige, C. K. H., and R. M. Huggins. 2018. "Inference from single occasion capture experiments using genetic markers." *Biometrical Journal* 60 (3): 463–479. https://doi.org/10.1002/bimj.201700046.

Humphrey, S. R. 1988. "Density estimates of the endangered Key Largo woodrat and cotton mouse (*Neotoma floridana smalli* and *Peromyscus gossypinus allapaticola*), using the nested-grid approach." *Journal of Mammalogy* 69 (3): 524–531.

Hurst, Z. M., R. A. McCleery, B. A. Collier, N. J. Silvy, P. J. Taylor, and A. Monadjem. 2014. "Linking changes in small mammal communities to ecosystem functions in an agricultural landscape." *Mammalian Biology* 79 (1): 17–23. https://doi.org/10.1016/j.mambio.2013.08.008.

Innes, D. G. L. 1978. "A reexamination of litter size in some North American microtines." *Canadian Journal of Zoology* 56 (7): 1488–1496.

Jett, D. A., and J. D. Nichols. 1987. "A field comparison of nested grid and trapping web density estimators." *Journal of Mammalogy* 68 (4): 888–892. https://doi.org/10.2307/1381576.

Johnsen, K., R. Boonstra, S. Boutin, O. Devineau, C. J. Krebs, and H. P. Andreassen. 2017. "Surviving winter: Food, but not habitat structure, prevents crashes in cyclic vole populations." *Ecology and Evolution* 7 (1): 115–124.

Jolly, G. M. 1965. "Explicit estimates from capture-recapture data with both death and immigration-stochastic model." *Biometrika* 52 (1/2): 225–247.

Jung, T. S., R. Boonstra, and C. J. Krebs. 2020. "Mark my words: Experts' choice of marking methods used in capture-mark-recapture studies of small mammals." *Journal of Mammalogy* 101 (1): 307–317. https://doi.org/10.1093/jmammal/gyz188.

Keiter, D. A., A. J. Davis, O. E. Rhodes Jr., F. L. Cunningham, J. C. Kilgo, K. M. Pepin, et al. 2017. "Effects of scale of movement, detection probability, and true population density on common methods of estimating population density." *Scientific Reports* 7. https://doi.org/10.1038/s41598-017-09746-5.

Kendall, W. L. 2001. "The robust design for capture-recapture studies: Analysis using Program MARK." In *Wildlife, Land, and People: Priorities for the 21st Century, Proceedings of the Second International Wildlife Management Congress*, edited by R. Field, R. J. Warren, H. Okarma, and P. R. Sievert, 357- 360. Bethesda: The Wildlife Society.

Kéry, M., and M. Schaub. 2012. *Bayesian Population Analysis Using WinBUGS: A Hierarchical Perspective*. Cambridge, MA: Academic Press.

King, K. L., J. A. Homyack, T. B. Wigley, D. A. Miller, and M. C. Kalcounis-Rueppell. 2014. "Response of rodent community structure and population demographics to intercropping switchgrass within loblolly pine plantations in a forest-dominated landscape." *Biomass & Bioenergy* 69: 255–264. https://doi.org/10.1016/j.biombioe.2014.07.006.

Kovacs, E. K., M. S. Crowther, J. K. Webb, and C. R. Dickman. 2012. "Population and behavioural responses of native prey to alien predation." *Oecologia* 168 (4): 947–957. https://doi.org/10.1007/s00442-011-2168-9.

Krebs, C. J. 1966. "Demographic changes in fluctuating populations of *Microtus californicus*." *Ecological Monographs* 36 (3): 239–273.

Krebs, C. J. 2014. *Ecological Methodology*. 3rd ed. Menlo Park, CA: Addison-Wesley Educational Publishers.

Krebs, C. J., R. Boonstra, S. Gilbert, D. Reid, A. J. Kenney, and E. J. Hofer. 2011. "Density estimation for small mammals from livetrapping grids: Rodents in northern Canada." *Journal of Mammalogy* 92 (5): 974–981. https://doi.org/10.1644/10-mamm-a-313.1.

Laake, J., and E. Rexstad. n.d. "RMark—an alternative approach to building linear models in MARK." In *Program MARK: A Gentle Introduction*, edited by E. Cooch and G. C. White, C1-C115. Accessed January 15, 2019. http://www.phidot.org/soft ware/mark/docs/book.

Laake, J. L., D. S. Johnson, and P. B. Conn. 2013. "marked: An R package for maximum likelihood and Markov Chain Monte Carlo analysis of capture-recapture data." *Methods in Ecology and Evolution* 4 (9): 885–890. https://doi.org/10.1111/2041-210x .12065.

La Haye, M. J. J., K. R. R. Swinnen, A. T. Kuiters, H. Leirs, and H. Siepel. 2014. "Modelling population dynamics of the common hamster (*Cricetus cricetus*): Timing of harvest as a critical aspect in the conservation of a highly endangered rodent." *Biological Conservation* 180: 53–61.

Lloyd, J. A., and J. J. Christian. 1969. "Reproductive activity of individual females in three experimental freely growing populations of house mice (*Mus musculus*)." *Journal of Mammalogy* 50 (1): 49–59. https://doi.org/10.2307/1378629.

Luikart, G., N. Ryman, D. A. Tallmon, M. K. Schwartz, and F. W. Allendorf. 2010. "Estimation of census and effective population sizes: The increasing usefulness of DNA-based approaches." *Conservation Genetics* 11: 355–373. https://doi.org/10.1007 /s10592-010-0050-7.

Lukacs, P. M., D. R. Anderson, and K. P. Burnham. 2005. "Evaluation of trapping-web designs." *Wildlife Research* 32 (2): 103–110. https://doi.org/10.1071/wr04011.

Lukacs, P. M., and K. P. Burnham. 2005. "Review of capture-recapture methods applicable to noninvasive genetic sampling." *Molecular Ecology* 14 (13): 3909–3919. https://doi.org/10.1111/j.1365-294X.2005.02717.x.

Mamba, M., N. J. Fasel, A. M. Themb'alilahlwa, J. D. Austin, R. A. McCleery, and A. Monadjem. 2019. "Influence of sugarcane plantations on the population dynamics and community structure of small mammals in a savanna-agricultural landscape." *Global Ecology and Conservation* 20 (2019): e00752.

Manly, B. F. J., S. C. Amstup, and T. L. McDonald. 2005. "Capture-recapture methods in practice." In *Handbook of Capture-Recapture Analysis*, edited by S. C. Amstrup, T. L. McDonald, and B. F. J. Manly, 58- 87. Princeton: Princeton University Press.

Mayfield, H. F. 1975. "Suggestions for calculating nest success." *Wilson Bulletin* 87 (4): 456–466.

McCleery, R. A. 2009. "Reproduction, juvenile survival and retention in an urban fox squirrel population." *Urban Ecosystems* 12 (2): 177–184.

McCleery, R. A., R. R. Lopez, N. J. Silvy, P. A. Frank, and S. B. Klett. 2006. "Population status and habitat selection of the endangered Key Largo woodrat." *The American Midland Naturalist* 155: 197–209.

McCleery, R.A., R. R. Lopez, N. J. Silvy, and D. L. Gallant. 2008. "Fox squirrel survival in urban and rural environments." *Journal of Wildlife Management* 72 (1): 133–137. https://doi.org/10.2193/2007-138.

McCleery, R. A., M. K. Oli, J. A. Hostetler, B. Karmacharya, D. Greene, C. Winchester, et al. 2013. "Are declines of an endangered mammal predation-driven, and can a captive-breeding and release program aid their recovery?" *Journal of Zoology* 291 (1): 59–68. https://doi.org/10.1111/jzo.12046.

McCleery, R. A., and I. D. Parker. 2011. "Influence of the urban environment on fox squirrel range overlap." *Journal of Zoology* 285 (3): 239–246. https://doi.org/10.1111 /j.1469-7998.2011.00835.x.

Menkens, G. E., and S. H. Anderson. 1988. "Estimation of small-mammal population-size." *Ecology* 69 (6): 1952–1959. https://doi.org/10.2307/1941172.

Miller, C. R., P. Joyce, and L. P. Waits. 2005. "A new method for estimating the size of small populations from genetic mark-recapture data." *Molecular Ecology* 14: 1991–2005. https://doi.org/10.1111/j.1365–294X.2005.02577.x.

Nichols, J. D. 2005. "Modern open-population capture-recapture models." In *Handbook of Capture-Recapture Analysis*, edited by S. C. Amstrup, T. L. McDonald, and B. F. J. Manly, 88–123. Princeton: Princeton University Press.

Nupp, T. E., and R. K. Swihart. 2000. "Landscape-level correlates of small-mammal assemblages in forest fragments of farmland." *Journal of Mammalogy* 81 (2): 512–526. https://doi.org/10.1644/1545–1542(2000)081<0512:llcosm>2.0.co;2.

Ostfeld, R. S., T. Levi, F. Keesing, K. Oggenfuss, and C. D. Canham. 2018. "Tick-borne disease risk in a forest food web." *Ecology* 99 (7): 1562–1573. https://doi.org/10.1002/ecy.2386.

Otis, D. L., K. P. Burnham, G. C. White, and D. R. Anderson. 1978. "Statistical-inference from capture data on closed animal populations." *Wildlife Monographs* 62: 1–135.

Palomares, F. 2001. "Comparison of 3 methods to estimate rabbit abundance in a Mediterranean environment." *Wildlife Society Bulletin* 29 (2): 578–585.

Parmenter, R. R., T. L. Yates, D. R. Anderson, K. P. Burnham, J. L. Dunnum, A. B. Franklin, et al. 2003. "Small-mammal density estimation: A field comparison of grid-based vs. web-based density estimators." *Ecological Monographs* 73 (1): 1–26. https://doi.org/10.1890/0012-9615(2003)073[0001:smdeaf]2.0.co;2.

Pennell, M. W., C. R. Stansbury, L. P. Waits, and C. R. Miller. 2013. "Capwire: A R package for estimating population census size from non-invasive genetic sampling." *Molecular Ecology Resources* 13 (1): 154–157. https://doi.org/10.1111/1755-0998.12019.

Petit, E., and N. Valière 2006. "Estimating population size with non-invasive capture-recapture data." *Conservation Biology* 20: 1062–1073.

Pierce, L. B., R. R. Lopez, and N. J. Silvy. 2012. "Estimating animal abundance." In *The Wildlife Techniques Manual*, vol. 1, 7th ed., edited by N. J. Silvy, 284–310. Baltimore: Johns Hopkins University Press.

Pinot, A., F. Barraquand, E. Tedesco, V. Lecoustre, V. Bretagnolle, and B. Gauffre. 2016. "Density-dependent reproduction causes winter crashes in a common vole population." *Population Ecology* 58 (3): 395–405. https://doi.org/10.1007/s10144-016-0552-3.

Pledger, S., K. H. Pollock, and J. L. Norris. 2003. "Open capture-recapture models with heterogeneity: I. Cormack-Jolly-Seber model." *Biometrics* 59 (4): 786–794. https://doi.org/10.1111/j.0006-341X.2003.00092.x.

Pledger, S., K. H. Pollock, and J. L. Norris. 2010. "Open capture-recapture models with heterogeneity: II. Jolly-Seber Model." *Biometrics* 66 (3): 883–890. https://doi.org/10.1111/j.1541-0420.2009.01361.x.

Pocock, M. J. O., A. C. Frantz, D. P. Cowan, P. C. L. White, and J. B. Searle. 2004. "Tapering bias inherent in minimum number alive (MNA) population indices." *Journal of Mammalogy* 85 (5): 959–962. https://doi.org/10.1644/bpr-023.

Pollock, K. H. 1982. "A capture-recapture design robust to unequal probability of capture." *Journal of Wildlife Management* 46 (3): 752–757. https://doi.org/10.2307/3808568.

Pollock, K. H., S. R. Winterstein, C. M. Bunck, and P. D. Curtis. 1989. "Survival analysis in telemetry studies: The staggered entry design." *Journal of Wildlife Management* 53 (1): 7–15. https://doi.org/10.2307/3801296.

Proença-Ferreira, A., C. Ferreira, I. Leitao, J. Pauperio, H. Sabino-Marques, S. Barbosa, et al. 2019. "Drivers of survival in a small mammal of conservation concern: An assessment using extensive genetic non-invasive sampling in fragmented farmland." *Biological Conservation* 230: 131–140. https://doi.org/10.1016/j.biocon.2018.12.021.

Rexstad, E., and K. P. Burnham. 1991. *User's Guide for Interactive Program CAPTURE: Color*. Fort Collins: Cooperative Fish and Wildlife Research Unit, Colorado State University.

Romairone, J., J. Jimenez, J. J. Luque-Larena, and F. Mougeot. 2018. "Spatial capture-recapture design and modelling for the study of small mammals." *PLoS ONE* 13 (6). https://doi.org/10.1371/journal.pone.0198766.

Rose, R. K. 2008. "Population ecology of the golden mouse." In *Golden Mouse: Ecology and Conservation*, edited by G. W. Barret and G. A. Feldhammer, 39–58. New York: Springer.

Rose, R. K., and J. W. Walke. 1988. "Seasonal use of nest boxes by *Peromyscus* and *Ochrotomys* in the dismal swamp of Virginia." *American Midland Naturalist* 120 (2): 258–267. https://doi.org/10.2307/2425997.

Ruibal, M., R. Peakall, A. Claridge, and K. Firestone. 2009. "Field-based evaluation of scat DNA methods to estimate population abundance of the spotted-tailed quoll (*Dasyurus maculatus*), a rare Australian marsupial." *Wildlife Research* 36: 721–736.

Ruiz-Capillas, P., C. Mata, and J. E. Malo. 2013. "Road verges are refuges for small mammal populations in extensively managed Mediterranean landscapes." *Biological Conservation* 158: 223–229. https://doi.org/10.1016/j.biocon.2012.09.025.

Schmidt, J. A., R. A. McCleery, P. M. Schmidt, N. J. Silvy, and R. R. Lopez. 2011. "Population estimation and monitoring of an endangered lagomorph." *The Journal of Wildlife Management* 75: 151–158.

Schnabel, Z. E. 1938. "The estimation of the total fish population of a lake." *The American Mathematical Monthly* 45 (6): 348–352.

Schumacher, F. X., and R. W. Eschmeyer. 1943. "The estimation of fish populations in lakes and ponds." *Journal of the Tennessee Academy of Science* 18: 228–249.

Schwarz, C. J., and A. N. Arnason. n.d. "Jolly-Seber models in MARK." In *Program MARK: A Gentle Introduction*, edited by E. Cooch and G. White, Chapter 12. Accessed July 20, 2019. http://www.phidot.org/software/mark/docs/book.

Seber, G. A. F. 1965. "A note on multiple-recapture census." *Biometrika* 52 (1/2): 249–259.

Seber, G. A. F. 1982. *The Estimation of Animal Abundance and Related Parameters*. New York: Macmillan Press.

Skalski, J. R., K. E. Ryding, and J. Millspaugh. 2005. *Wildlife Demography: Analysis of Sex, Age, and Count Data*. Burlington, MA: Elsevier.

Smith, W. P. 2012. "Flying squirrel demography varies between island communities with and without red squirrels." *Northwest Science* 86 (1): 27–38. https://doi.org/10.3955/046.086.0103.

Sotola, V. A., and D. E. Garneau. 2014. "Survey of the patterns of nest box use among squirrels (Sciuridae) in managed forest stands in Clinton County, New York." *Open Ecology Journal* 7: 1–8.

Spitzer, R., A. J. Norman, M. Schneider, and G. Spong. 2016. "Estimating population size using single-nucleotide polymorphism-based pedigree data." *Ecology and Evolution* 6: 3174–3184.

Sullivan, T. P., and D. S. Sullivan. 2017. "Old-growth characteristics 20 years after thinning and repeated fertilization of lodgepole pine forest: Tree growth, structural attributes, and red-backed voles." *Forest Ecology and Management* 391: 207–220. https://doi.org/10.1016/j.foreco.2017.02.021.

Sullivan, T. P., D. S. Sullivan, R. Boonstra, C. J. Krebs, and A. Vyse. 2017. "Mechanisms of population limitation in the southern red-backed vole in conifer forests of western North America: Insights from a long-term study." *Journal of Mammalogy* 98 (5): 1367–1378. https://doi.org/10.1093/jmammal/gyx082.

Tanaka, R. 1972. "Investigation into the edge effect by use of capture-recapture data in a vole population." *Researches on Population Ecology* 13 (2): 127–151. https://doi.org/10.1007/bf02521973.

Topp-Jorgensen, J. E., A. R. Marshal, H. Brink, and U. B. Pedersen. 2008. "Quantifying the response of tree hyraxes (*Dendrohyrax validus*) to human disturbance in the Udzungwa Mountains, Tanzania." *Tropical Conservation Science* 1 (1): 63–74. https://doi.org/10.1177/194008290800100106.

Wangdwei, M., B. Steele, and R. B. Harris. 2013. "Demographic responses of plateau pikas to vegetation cover and land use in the Tibet Autonomous Region, China." *Journal of Mammalogy* 94 (5): 1077–1086. https://doi.org/10.1644/12-mamm-a-253.1.

Warren, A. E., L. M. Conner, S. B. Castleberry, and D. Markewitz. 2017. "Home range, survival, and activity patterns of the southeastern pocket gopher: Implications for translocation." *Journal of Fish and Wildlife Management* 8 (2): 544–557. https://doi.org/10.3996/032017-jfwm-023.

Wayne, A. F., M. A. Maxwell, C. G. Ward, C. V. Vellios, M. R. Williams, and K. H. Pollock. 2016. "The responses of a critically endangered mycophagous marsupial (*Bettongia penicillata*) to timber harvesting in a native eucalypt forest." *Forest Ecology and Management* 363: 190–199. https://doi.org/10.1016/j.foreco.2015.12.019.

White, G. C., and Kenneth P. Burnham. 1999. "Program MARK: Survival estimation from populations of marked animals." *Bird Study* 46 (Suppl.): S120–S139.

White, G. C., K. P. Burnham, D. L. Otis, and D. R. Anderson. 1978. *User's Manual for Program CAPTURE*. Logan UT: Utah State University Press.

Williams, B. K., J. D. Nichols, and M. J. Conroy. 2002. *Analysis and Management of Animal Populations*. San Diego: Academic Press.

Wilson, K. R., and D. R. Anderson. 1985. "Evaluation of a density estimator based on a trapping web and distance sampling theory." *Ecology* 66 (4): 1185–1194. https://doi.org/10.2307/1939171.

Young, H. S., R. Dirzo, K. M. Helgen, D. J. McCauley, S. A. Billeter, M. Y. Kosoy, et al. 2014. "Declines in large wildlife increase landscape-level prevalence of rodent-borne disease in Africa." *Proceedings of the National Academy of Sciences of the United States of America* 111 (19): 7036–7041. https://doi.org/10.1073/pnas.1404958111.

Zippin, C. 1958. "The removal method of population estimation." *Journal of Wildlife Management* 22 (1): 82–90. https://doi.org/10.2307/3797301.

human visitation (Stephenson 1993). However, most of these correlative studies use a space-for-time replacement approach, using spatial changes to understand how human alterations have changed communities over time.

Correlating changes in small mammal assemblages with changes in the environment over time can be laborious, but it has been foundational in our understanding of small mammal ecology. Long-term data sets (>7 years) have been used to show the response of small mammal assemblages to both pulsed and episodic rainfall (Whitford 1976, Brown and Heske 1990, Leirs et al. 1996) and changes in climate (Myers et al. 2009). Similarly, datasets have linked the temporal changes in small mammal assemblages with masting events (Schnurr et al. 2002, Clotfelter et al. 2007), plant cover (Ernest et al. 2000), fire (Monadjem and Perrin 2003), and successional stage (Fox 1990). Collecting data over time has also allowed researchers to understand population cycling (Korpimaki et al. 2005), disease and predator-prey dynamics (Hanski et al. 1993) and species interactions (Swihart and Slade 1990). In fact, after collecting 15 years of data, Swihart and Slade (1990) concluded that they would have reached a different conclusion had they only studied their small mammal assemblages for a more conventional period of two to three years. Collecting long-term data has been an invaluable tool for understanding small mammal communities, but it does not allow researchers to isolate the causal mechanisms that drive community change.

Experimental Approaches

Few, if any, vertebrate taxonomic groups lend themselves to experiments in the way small mammal communities do. While there are entire fields dedicated to behavioral and physiological experiments of small mammals, here we are interested in presenting the different ways that researchers have manipulated the environment to study how small mammal communities change and interact with their surroundings. Manipulating important components of the environment to understand the mechanisms that shape small mammal communities has a rich and important history in ecology. One approach for manipulating small mammal communities is to use a field enclosure (Fig. 10.1). Field enclosures commonly use buried fencing to prevent immigration and emigration of small mammals. Fences can be constructed in a way that either allows or discourages predators from entering the area (Darracq et al. 2016). Small mammals are measured or stocked, and treatments and controls are randomly assigned. These types of structures have been used to understand the influence of dispersal (Krebs et al. 1973), fire (Crowner and Barrett 1979), and competition (Eccard and Ylonen 2003) on small mammal assemblages. Fenced enclosures can also be used to isolate the influence of small

Fig. 10.1. Small mammal enclosure at The Jones Center at Ichauway, Georgia

mammal communities on vegetation (Krebs et al. 1969), seeds (Ostfeld et al. 1997), and other components of the environment.

Ecologists have used fencing to exclude small mammals from some research plots while allowing them to access others. Researchers have used this approach to understand the influence of small mammals on seed survival (Bricker et al. 2010) and invertebrate communities (Churchfield et al. 1991). Researchers have also used wire-mesh fencing to exclude larger mammals

Fig. 10.2. Large mammal exclusion fence at Kruger National Park, South Africa

from research plots while allowing small mammals to move freely on and off the plots (Fig. 10.2). This type of experiment has been effective at investigating the interactions between small mammal communities and large mammalian browsers and grazers (Keesing 1998, Flowerdew and Ellwood 2001, Saetnan and Skarpe 2006, Torre et al. 2007) as well as predation (Korpimaki et al. 2004).

In addition to fences, researchers have used an endless variety of manipulations to understand relationships between small mammal communities and the environment. Researchers have added food, water, nitrogen, and other supplements to influence bottom-up processes (Grant et al. 1977, Abramsky et al. 1979, Meserve et al. 2001, G. Morris et al. 2011) and removed predators to understand top-down processes (Korpimaki and Norrdahl 1998, Risbey et al. 2000). Researchers have removed small mammals from their study sites to investigate species interactions (Brown and Munger 1985). For example, we recently conducted a study to understand competitive interactions of fox squirrels (*Sciurus niger*) and gray squirrels (*Sciurus carolinensis*) by translocating gray squirrels out of the small patches of hardwood forest that they appear to dominate (Sovie et al. 2021). Small mammal communities have also made excellent models for studying land management practices and manipulations such as logging (Moses and Boutin 2001), prescribed fire (Andersen et al. 2005, G. Morris et al. 2011), harvesting of agricultural fields (Mamba et al. 2019), and invasive species management (Harris 2009).

Finally, although they are rarely utilized, experiments that manipulate systems for long periods can help disentangle the complex relationships of small mammal populations on their environment. Possibly the most notable example of this is a long-term experiment that has been conducted by Dr. James Brown, his colleagues, and his students in the Chihuahuan Desert (Brown et al. 2001). This experiment included the removal different rodent species from research plots coupled with the continued monitoring of small mammals, plants, ants, and weather. This research has provided a continued source of information on the complex nonlinear relationships between climate change and rainfall, plant growth, seed predation, and rodents (Brown et al. 2001, Ernest et al. 2016).

Quantifying Communities

Biodiversity

Most studies of small mammal communities include a measure of biodiversity. However, the difficulty with this seemingly basic metric is that it can be measured and conceptualized in a myriad of different ways. For example, the variety of life can be measured based on species, morphology, subspecies, functional groups, genetic variation, or evolutionary history. Selection among these approaches should be informed by the questions and problems researchers are hoping to address with their work. Yet, even after deciding on the particular measures of biodiversity, classifying taxonomy (Chapter 6) and particular functional groups or traits of animals can be equally challenging. Our goal here is not to delve into the nuances of conceptualizing and measuring biodiversity (see Magurran and McGill 2011 for a comprehensive examination of biodiversity), instead we aim to present the broad concepts and techniques that are applicable to the study of small mammal communities.

Species Richness

The number of species in a defined area or community, species richness, is the oldest and most straightforward approach to quantifying diversity of animal communities. Nonetheless, we know that the number and types of small mammals detected in specious communities will be influenced by researcher effort and detection technique (Bovendorp et al. 2017), so caution should be used when comparing these measures across study designs. If researchers are interested in understanding how environmental gradients (e.g., elevation, disturbance, canopy cover, patch size), land uses (e.g., urban, agricultural, rural), or treatments (e.g., fire, predator removal, fertilizer) alter the number of small mammal species, then using consistent methods and efforts with raw species counts can yield biologically meaningful results. If different methodologies or efforts are used, it is possible to adjust for these differences (see Bovendorp et al. 2017) or attempt to estimate the actual number of species in the community. There are three common approaches to estimating species richness:

(1) nonparametric estimators based on frequencies that use information on rare species to estimate undetected species, (2) using species accumulation curves to estimate the number of species not detected, and (3) using species detection probabilities to estimate the number of species missed.

Nonparametric Estimators

Some frequently used nonparametric estimators for small mammal community richness are the jackknife (Jack) (McCain 2004), Chao (Chao 2005), and convergence-based estimators (Moritz et al. 2008). Jackknife estimators determine the number of species missed, using presence/absence or abundance data, to determine the amount of unique information from species found in only one (*Jack 1*) or two (*Jack 2*) samples (Chao 2005). The more sampling plots with unique species—in either one (*Jack 1*) or two (*Jack 2*)—the more missed species are assumed and the greater the difference in the jackknife estimate of richness and observed species richness. Similar to the jackknife estimator, the Chao estimator uses information on rare species (found in only one or two samples) to estimate the number of species that might have been missed (Chao 2005). The *Chao 1* estimator can be used on data that includes the number of small mammals in a sample, while the *Chao 2* estimator works with presence/absence data. Convergence estimators also use information on rare species to estimate species richness. These estimators measure the number of undetected species using the number of rare species and the number of species detected only once (Chao 2005). To determine if a species is rare, the abundance convergence estimator (ACE) often uses a cutoff of 10 individuals, while the indices (presence/absence) convergence estimator (ICE) often uses a cutoff based on presence in 10 samples (Chao and Chiu 2016).

We were unable to find an evaluation of these nonparametric estimators of richness applied to small mammals; however, all the estimators presented here appear to be useful and can be used to compare richness estimates derived from different sampling strategies (Walther and Moore 2005, Hortal et al. 2006). In particular, the ACE, *Chao 1*, and *Jack 1* and *2* all appear to perform well as estimators, regardless of the scale or gain the samples are used to estimate (Hortal et al. 2006). Furthermore, a meta-analysis of the accuracy and bias of 10 estimators found the *Chao 1* and *Jack 1* and *2* outperformed other metrics. Other research suggested the convergence-based estimators (ICE, ACE) provide good results with large samples (Chazdon et al. 1998, Hortal et al. 2006, Chao and Chiu 2016). Alternatively, while the jackknife estimators perform well in many studies, Chao and Chiu (2016) point out that these estimators likely underestimate true richness for small samples and overestimate true richness for large samples. More practically, when researchers estimate the richness of small mammal communities, it is not uncommon for them to use several estimators that all provide generally similar results (McCain 2004, Moritz et al. 2008, Sanchez-Giraldo and Diaz-Nieto 2015). Most of these estimates can be calculated using the vegan package in the R statistical platform or with the EstimateS program developed by Robert Colwell (http://viceroy.eeb.uconn.edu/estimates/) (Table 10.1). To get the most out of the vegan package, see the comprehensive guide by Gardener (2014).

Species Accumulation Curves

Another option for comparing species richness across communities with different sampling efforts is to use species accumulation curves that predict the number of species detected as a function of sampling effort but not methodologies. This is commonly achieved using rarefaction, which estimates the number of species expected in a random sample of individuals. By randomly pulling samples of individuals from a community with a greater number of samples, we can determine if the species richness of that community differs from that of a community with a smaller number of samples (Gotelli and Colwell 2011). In addition to point estimates, this random subsampling can be used to construct rarefaction curves that estimate species richness across the range of sample sizes (Fig. 10.3). Rarefaction can be computed with random subsamples that come from the entire sample of individuals observed (individual-base rarefaction) or from the number of species taken at random (sample-base rarefaction) from sets of samples (e.g., trapping grids). Sample-based rarefaction can use both presence absence and abundance data, while individual-base rarefaction can only use data that includes the number of animals. Gotelli and Colwell (2011) suggest that the sample-based rarefaction is better for most ecological studies because it maintains the independence of most sampling units, so data from different trapping grids or habitat types are not aggregated.

Ecologists studying small mammals often use rarefaction and accumulation curves to determine if their sampling has reached an asymptote (Goodman and Rakotondravony 2000, Nor 2001, Rickart et al. 2011). Here, they assume that if no more species are being added with addition of more samples, then estimated species richness is reflective of actual species richness. However, several authors argue against the use of rarefaction curves to estimate species richness in specious communities and stress that it should not be extrapolated for estimating species richness for areas larger than the area sampled (Gotelli and Colwell 2011, Krebs 2014).

Occupancy Modeling

One of the problems with collecting data on communities is that species are almost always detected at different rates, and some species are not detected

Table 10.1. Commonly used software for the analysis of small mammal communities. Software is grouped by R-based packages and other software. The table provides information of costs, URLs, and important features.

Software	Free	URL	Important Features
R packages			
vegan	Yes	https://cran.r-project.org/web/packages/vegan/vegan.pdf	Diversity metrics, evenness, richness estimators, accumulation curves, similarity metrics, ordination, NMMDS, ANOSIM, PERMANOVA
betapart	Yes	https://cran.r-project.org/web/packages/betapart/betapart.pdf	Partitions turnover and nestedness
ade4	Yes	https://cran.r-project.org/web/packages/ade4/ade4.pdf	Ordination, redundancy analysis
labdsv	Yes	https://cran.r-project.org/web/packages/labdsv/labdsv.pdf	Ordination, generalized additive model (GAM) against ordination
iNext	Yes	https://cran.r-project.org/web/packages/iNEXT/iNEXT.pdf	Hill numbers
FD	Yes	https://cran.r-project.org/web/packages/FD/FD.pdf	Functional diversity metrics
picante	Yes	https://cran.r-project.org/web/packages/picante/picante.pdf	Phylogenetic diversity metrics
Other Software			
EstimateS	Yes	http://viceroy.eeb.uconn.edu/estimates/EstimateSPages/AboutEstimateS.htm	Diversity metrics, richness estimators, accumulation curves, similarity metrics
Pisces, Species Diversity & Richness (SDR)	No	http://www.pisces-conservation.com/softdiversity.html	Diversity metrics, evenness, richness estimators, accumulation curves
Pisces, Community Analysis Package (CAP)	No	http://www.pisces-conservation.com/softcap.html	Ordination, similarity metrics, NMMDS, ANOSIM
Primer	No	https://www.primer-e.com/our-software/primer-version-7/	Ordination, similarity metrics, NMMDS, ANOSIM, functional diversity metrics
PC-ORD	No	https://www.wildblueberrymedia.net/pcord	Ordination, similarity metrics, NMMDS, PERMANOVA, functional diversity metrics
CANOCO	No	http://www.canoco5.com/	Ordination, similarity metrics, NMMDS
MVSP	No	https://www.kovcomp.co.uk/mvsp/mvsp3fea.htmL	Diversity metrics, similarity metrics, ordination
PAST	Yes	https://palaeo-electronica.org/2001_1/past/issue1_01.htm	Ordination, multivariate analyses

Note: ANOSIM = analysis of similarity; PERMANOVA = permutational multivariate analysis of variance; NMMDS = nonmetric multidimensional scaling.

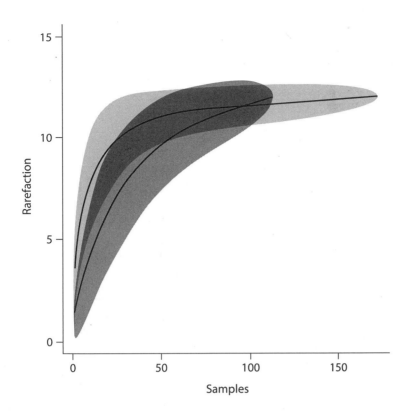

Fig. 10.3. Rarefaction estimates and confidence intervals as a function of the number of sampling plots of two populations of small mammals

at all (MacKenzie et al. 2017). Accordingly, multispecies occupancy models were developed to adjust presence/absence data based on species-level detection probabilities. To deal with the strong possibility that some species go undetected, researchers may augment multispecies occupancy models with a predetermined number of potentially missed species represented by 0s (Kéry and Royle 2015). Then, together with detected species, a Bayesian hierarchical approach is used to generate a detection-error corrected estimate of species richness (species presence/absence) for each site (Kéry and Royle 2015). There are several advantages to this approach relative to the nonparametric and species rarefaction approaches. Community occupancy modeling allows you to incorporate covariates that can inform the detection and occupancy of species. This information allows you to extend your inference beyond the areas sampled. Additionally, because the community occupancy approach models species responses, it allows you understand how species-level responses shape community metrics. For example, Loggins et al. (2019) modeled the response of small mammal species across grass and woody cover gradients. They showed a weak but consistent response of species to grass that lead to a positive community response. Alternatively, they showed stronger but mixed (positive and negative) species responses to woody cover

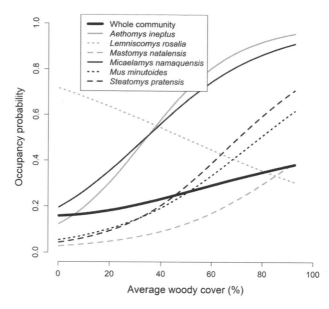

Fig. 10.4. The modeled occupancy probability for small mammal species along a woody cover gradient. The average community effect is displayed as a thick black line. *(Modified from data presented in Loggins et al. 2019)*

that did not produce a biologically relevant community response (Fig. 10.4). While it has been more common for researchers to model species-specific occupancy of small mammals (Kalies et al. 2012, Rowe et al. 2015, Faller and McCleery 2017, Monadjem et al. 2018), we believe there are clear advantages to using multispecies occupancy modeling to estimate species richness and other community parameters. Potential drawbacks to these models are that they can be computationally challenging and, because they only use presence/absence data, important abundance data cannot be used.

Diversity

The concept of species diversity differs from richness because it combines species-richness information with information on species abundances (Maurer and McGill 2011). Species diversity is a function of evenness, or the differences in the relative abundances of species in a community. Ecologists have a long history of developing measures of diversity that include measures of richness and evenness (Gotelli and Chao 2013). Our goal here is to present and evaluate the more common methods that have been used in the research of small mammal communities. For a more complete account of diversity metrics, we recommend Maurer and McGill (2011). Additionally, we focus on the use of diversity metrics for the estimation of species-level diversity. We touch on the use of diversity metrics used to estimate taxonomic and functional diversity later in the chapter.

Due to computational complexity, assumption about distributions, and a lack of theoretical underpinning for parametric approaches for measuring diversity, the most commonly used metrics to estimate species diversity are nonparametric (Krebs 2014). One of the most common nonparametric indices used to estimate the diversity of small mammal communities is Shannon's Index (also known as Shannon-Weiner or Shannon entropy). This index is based on information theory that attempts to quantify the uncertainty in the species identity of randomly chosen individuals (Gotelli and Colwell 2011). In diverse systems (more uncertainty), a random individual could be from any number of species, but in a simplified system (less uncertainty), it would be easier to predict an individual's species from a few potential options. The greater the diversity, the larger the uncertainty, the greater the measure of Shannon's diversity that effectively ranges from 0–5 (Krebs 2014).

The first nonparametric method that was developed for estimating species diversity that is still widely used is Simpson's Index. The idea behind this measure of diversity is that diversity is inversely related to the probability that two randomly picked individuals belong to the same species from an infinite sample population (Krebs 2014). The original formulation of the Simpson's Index has since been adjusted (Inverse Simpson's Index) to increase with increasing diversity (McCune and Grace 2002) and to account for the total number of individuals in the assemblage (Gini-Simpson Index). This Gini-Simpson Index can be easily interpreted as the probability that two randomly chosen individuals are different species (Gotelli and Chao 2013). Finally, using the proportion of each species in the community, MacArthur (1965) transformed the Gini-Simpson Index to units of species numbers. This new index, known a "Hill Numbers" after ecologist Mark Hill, is preferred because it is easily interpreted as the number of equally abundant species, where if all species were equally abundant, the index would equal species richness in the sample (Gotelli and Chao 2013, Krebs 2014). However, the use of Hill Numbers is exceedingly rare in the ecological study of small mammals (but see de la Sancha 2014).

Simpson's (Briani et al. 2004, Suzan et al. 2009, Hurst et al. 2013) and Shannon's (Maisonneuve and Rioux 2001, Michel et al. 2006, Heroldova et al. 2007) indices are both commonly used in the study of small mammal ecology, and it is not uncommon for researchers to use both metrics, but they do perform differently. Simpson's Index performs well with limited sampling of a target community but can be biased toward abundant species (Hill et al. 2003). Looking at plants and arthropods, E. K. Morris et al. (2014) found that Simpson's Index performed better than Shannon's when detecting differences at a fine-scale site level. With extensive sampling, Shannon's Index is a generally useful measure of richness and evenness, but it can underestimate diversity, particularly with limited samples of the community, because it gives more weight per individual to rare species, and it can be difficult to interpret because it is not on a biologically meaningful scale (Hill et al. 2003).

Evenness and Dominance

Diversity measures attempt to combine species richness and evenness of a community. As you saw above, there are a number of ways to estimate species richness independent of evenness. Similarly, sometimes small mammal ecologists have been interested in understanding the evenness, equitability, or the relative abundance of species within a community, irrespective of species richness (Kirkland 1990, Miller et al. 2004, Ostoja and Schupp 2009). Ecologist have used numerous methods to quantify evenness (see Smith and Wilson 1996), but two common metrics, Shannon's and Simpson's diversity indices, can be adjusted to eliminate the influence of species richness, producing Shannon's evenness and Simpson's evenness (Maurer and McGill 2011). Two other measures of evenness that small mammal researchers have used and that appear to be relatively independent of richness are Smith Wilson evenness, E_{var} (Avenant and Cavallini 2007), and Camargo evenness (Dreiss et al. 2015). After evaluating the performance of evenness measure, Smith and Wilson (1996) suggested their own metrics based on the variance of scaled logged abundance of the species in the community. Alternatively, Camargo's evenness is a measure of the deviation for the highest possible evenness where all species have the same abundance (Maurer and McGill 2011, Krebs 2014).

Analogous to evenness are measures of dominance that quantify how much one or several species dominate a community. Many small mammal communities vary in the numerical dominance of species over time and space (Anthony et al. 1981, Heroldova et al. 2007, Blois et al. 2010); thus, it is easy to see how metrics of dominance could be useful for describing small mammal communities. However, previous research has been more inclined to use measures of evenness to capture these types of changes in and across small mammal assemblages. Relative dominance, or Berger-Parker indices of the most common species in an assemblage, can be measured as the abundance of that species divided by the total abundance in the assemblage (Maurer and McGill 2011). Additionally, there are metrics based the scaled proportional abundance of the two (McNaughton and Wolf 1970) or three (Misra and Misra 1981) most common species.

Composition/Beta Diversity

Biodiversity is often conceptualized and measured spatially. The metrics of biodiversity that we have presented thus far have generally been measures of alpha diversity, diversity of defined local areas or sampling plots. This is just one component of diversity that has been conceptualized in three interrelated components of regional diversity (gamma diversity), which is composed of local alpha diversity and variation in the diversity and composition between these areas, or beta diversity (Whittaker 1972, McCune and Grace. 2002, Jost et al. 2011). Understanding differences and similarities in species composition among communities and across ecological gradients is fundamental to advancing ecology. Small mammal ecologists have long been interested in understanding if communities are different and determining what features or processes are causing those differences. Accordingly, we broadly consider measures of the differences and similarities of communities' composition to be measures of beta diversity.

Similarity

The earliest approaches to examine compositional similarity of assemblages were to use presence/absence-based indices. Two of the oldest and most commonly used indices are the Jaccard Index and Sørensen Index. Small mammal researchers have used these metrics to understand changes between small mammal assemblages over time and across ecological gradients (de la Pena et al. 2003, Burel et al. 2004, Blois et al. 2010, Dambros et al. 2015). The Jaccard Index is calculated by comparing the number of shared species to the total number of species in all samples (Gotelli and Chao 2013). The Sørensen Index is calculated by comparing the number of similar species in assemblages to the average number of species in all assemblages. Thus, the two metrics differ in their scale of interest, broad (Jaccard) and local (Sørensen). Both metrics were originally designed to compare only two assemblages, but they have been modified to compare similarities across multiple assemblages (Jost et al. 2011). While both estimators are broadly used and easily interpretable, they are likely to underestimate similarity because they do not account for undetected species that are shared between samples (Gotelli and Chao 2013). The best ways of dealing with this potential bias are to sample assemblages until species richness becomes asymptotic or to use density-based indices of similarity.

A more ecologically sound approach to evaluating the amount of similarity between assemblages is to use abundance-based indices which evaluate different species frequencies among communities (Jost et al. 2011). There is no shortage of indices to choose from (see Jost et al. 2011), but from an informal review of small mammal research (Google Scholar and Web of Science), it is clear the Bray-Curtis similarities index (0 = similar, 1 = dissimilar) is most commonly used and that it is an effective measure of abundance-based similarity (McCune and Grace 2002). The Bray-Curtis Index was derived from the Sørensen Index, and if species in each assemblage are equally abundant, it is effectively a Sørensen Index (Jost et al. 2011). However, several authors have pointed out that the index uses absolute abundances, confounding similarity and density, where assemblages with equal ratios of species but different absolute numbers are not equal (Chao et al. 2005, Jost et al. 2011). One option is to standardize data before using this index (Clarke et al. 2014) or using alternatives such as the Morista-Horn and Renkonen indices, which are not affected by sample size and absolute abundance of assemblages (Jost et al. 2011, Krebs 2014).

Ordination

Another approach to quantifying differences among assemblages is ordination, which allows ecologists to map assemblages across ecological gradients and in multidimensional space. Ordination attempts to reflect the difference in assemblages as distances, where assemblages that are similar are located close together and assemblages that are dissimilar are farther apart. The original eigenvector ordination technique, principal component analysis (PCA), uses raw data of species abundance and maximizes the variation between assemblages by fitting multiple lines (or components) projected on two (or more) dimensional space (Clarke et. al 2014). For this approach to be successful, a large portion of the variability in the assemblages must be captured in a few dimensions; however, in communities with lots of species, it is possible that only a small portion of the variation will be captured with only the two components needed for a two-dimensional plot (Clarke et al. 2014). The original PCA has been used to understand differences in small mammal assemblages (Gitzen et al. 2007, Michel et al. 2007), but a drawback is that it requires multivariate normality and data with roughly linear relationships and, unfortunately, this is rarely the case of community composition data (McCune and Grace 2002). Principle coordinate analysis is a variant of PCA that is more flexible and improves performance with community data. Other similar ordination techniques used in small mammal research include detrended correspondence analysis (DCA) and noncentered PCA. Both methods are recommended in specific instances but still have limitations (see McCune and Grace 2002, Clarke et al. 2014). One of the advantages of PCA and other eigenvector ordination techniques is that they can provide useful information on the organization of the community. They also provide eigenvalues, which represent the variation explained in each component, and factor loadings describing the strength of the relationship (analogous to Persons r^2) between components and the data (species in this instance). In this manner, researchers can determine which species are shaping the composition of assemblages in a community (see Gitzen et al. 2007).

A broadly recommended alternative to PCA and its variants is to map similarities across assemblages is using nonmetric multidimensional scaling

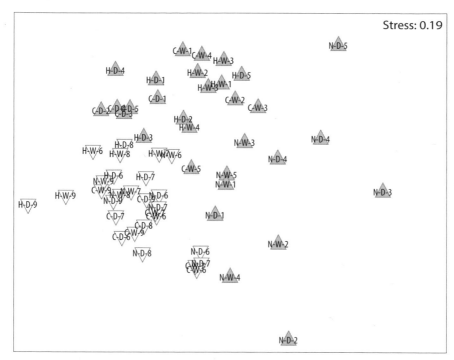

Fig. 10.5. Example of nonmetric multi-dimensional scaling (NMMDS) showing the similarity of communities found within the conservation area (shaded triangles) and sugarcane area (white triangles) *(Modified from Hurst et al. 2013)*

(NMMDS) (McCune and Grace 2002, Clarke et al. 2014). The advantage of NMMDS is that it can be used with non-normal data, does not assume linear relationships among variables, and unlike other methods, can use data with many zeros (McCune and Grace 2002). Instead of using species abundances, NMMDS uses a similarity matrix based on the similarity indices, like the ones presented above or others. Based on this similarity matrix, NMMDS then maps similarities based on the ranking among assemblages, where assemblages that are ranked farther apart will be farther apart on the map (Clarke et al. 2014). In examining the influence of agriculture on small mammals in Eswatini, Hurst et al. (2013) used a Bray-Curtis similarity matrix and NMMDS to show the clear distinction in small mammal communities between sugarcane and conservation areas (Fig. 10.5). The drawback to this approach is that it is difficult to derive any additional information about the relations among the communities, such as with the factor loadings provided in PCA.

Linking Communities to Environmental Variation

Ordination by itself can describe the relationship among assemblages, and you can label different properties of assemblages (e.g., site, year) to understand how they vary. However, most ecologists are interested in understanding whether communities vary as a function of environmental or temporal variation. There are several different approaches to differentiate between assemblages that can be grouped into categories such as year, site, season, and land use. Both the permutational multivariate analysis of variance (PER-MANOVA) procedure and analysis of similarity (ANOSIM) are commonly used for analysis-of-variance-like tests based on similarity (dissimilarity or distance) matrices that compare groups (Anderson and Walsh 2013) of small mammal assemblages (see Stephens et al. 2017, Simelane et al. 2018). However, the PERMANOVA procedure appears to be more powerful at detecting differences in community structure and more robust to heterogeneity (Anderson and Walsh 2013). Another approach comparing small mammal assemblages is discriminant analysis, which can be used to determine if assemblages or other multivariate data (e.g., vegetation) can be classified into predefined groups. This approach has been used to determine if small mammal assemblages differ among sites (Michel et al. 2006), but small mammal researchers have more commonly used it to determine if species can be classified based on microhabitat variables that they use (see Dueser and Shugart 1978).

As most environmental data are continuous (e.g., rainfall, canopy cover, salinity), ecologists have also developed a number of approaches for relating species assemblages to this type of data. The preferred approach has been constrained ordination or direct gradient analysis. The commonly used canonical correspondence analysis (CCA) constrains the ordination of species abundance by environmental variables so that the ordination axes for the assemblages are described by the environmental gradients of interest (McGarigal et al. 2000). A variant of CCA called redundancy analysis (RDA) uses a linear model to calculate the amount of variance in community data explained by the environmental data (McGarigal et al. 2000). An extension of RDA that allows users to choose among similarity or distance matrices (i.e., Bray-Curtis) is called distance-based redundancy analysis (Legendre and Anderson 1999). Analogous to the inclusion of random effects, this method was modified to allow users to block or remove the variance from some variables (McArdle and Anderson 2001). The data for constrained ordinations are typically presented in the form of an ordination diagram where the principal components are (1) the axes, (2) the lengths of the arrows, which are used to represent variable strength, and (3) the arrow directions, which indicate the positive or negative relationship with the axes (Fig. 10.6). Outputs include the portion of variance in the community data explained by each environmental variable. Additionally, you can further test the significance each variable using permutation tests (Legendre and Anderson 1999). It is important to caution that there can be several issues with the assumptions of parametric multivariate methods (e.g., normalcy, linear relationships, no collinearity) used in these analyses. Further, large numbers of environmental variables relative to the number of observations can lead to questionable results (McCune and Grace 2002).

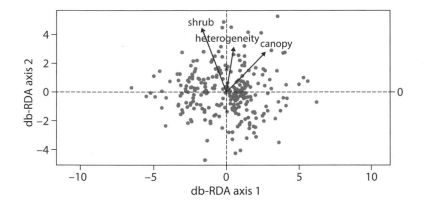

Fig. 10.6. Distance-based redundancy (db-RDA) plots showing the relationship between small mammal communities and shrub cover, canopy cover, and heterogeneity index *(Modified from McCleery et al. 2018)*

Turnover and Nestedness

Until this point we have considered the variation or beta diversity among communities without focusing on the processes that create that variation. To better understand these processes, we can conceptualize the difference in communities as caused by nestedness and turnover (Baselga 2010). Turnover is the replacement of some species by others, while nestedness is a pattern where change in similarity among sites comes from the loss of species from the most specious site (Baselga et al. 2017). Accordingly, there are several metrics for abundance and indices (presence/absence) that can be used to independently calculate or partition the similarity of assemblages from turnover and nestedness (see Baselga 2010). For example, both the Jaccard Index (Dambros et al. 2015) and Sørensen Index (Beca et al. 2017) have been used to partition similarity into turnover and nestedness to understand the response of small mammal communities to land alteration and climate in the Atlantic Forest in Brazil.

Functional and Phylogenic Composition

We have broadly addressed ways to quantify communities of small mammals based on a species concept; however, there has been a growing interest in measuring communities based on their function and phylogenies. Another way to conceptualize a small mammal community is as a collection of functional traits that are phenotypic characteristics that influence species fitness and/or ecosystem processes (Weiher 2011). You can think of the functional traits of small mammals as things, like mass, diet (e.g., % granivory, % insectivory), and activity periods (e.g., nocturnal, diurnal), that can be collected from field data, field guides, or readily available databases, such as the species-level foraging attributes of the world's birds and mammals (Wilman et al. 2014). Just like species, these traits can then be used to describe communities in terms of diversity, evenness, similarity, and other community measures.

Functional diversity is commonly measured as functional dispersion (FDis) because it tends to be uncorrelated with species richness and it can be calculated using any distance or dissimilarity measure (Laliberté and Legendre 2010). For example, McCleery et al. (2018) showed that FDis of small mammals changed across a gradient of woody cover in southern Africa. Analogous and often complementary to using traits to understand communities is the use of phylogenic information to explore diversity. All measures of phylogenetic diversity need to start with a phylogeny of the species in the assemblage(s) of interest (Vellend et al. 2011), often populated from information from a public genetics database (e.g., GenBank, National Center for Biotechnology Information). From this information a measure of distinctness can be generated for each species and then summed or averaged for the assemblage (Vellend et al. 2011). Small mammal ecologists have only begun to explain communities based on functional and phylogenic traits (Dreiss et al. 2015, Luza et al. 2015, Bovendorp et al. 2019, Garcia-Navas 2019), but this is an area of research with considerable potential for conservation because of its ability to describe evolutionary diversity and ecologically relevant traits of small mammal communities and its ability to provide considerably more information than species-based measures of community heterogeneity alone (Brum et al. 2017).

Conclusions and Future Directions

The study of small mammal communities through field surveys and experiments has been foundational to our broader understanding of ecology and conservation biology. Researchers have used a broad array of quantitative approaches to examine variation and changes in small mammal assemblages, but species richness, diversity, and evenness are the most commonly used metrics. Most of these metrics can be adjusted to account for species that were not detected during sampling; however, comparing adjusted and unadjusted community metrics across sampling regimes may have inherent biases. As ecologists increasingly look to aggregate data and examine ecological patterns across scales, we expect to see an increasing effort to adjust or standardize small mammal community datasets. Accordingly, we expect to see a greater use of multispecies occupancy and other Bayesian hierarchal approaches to model small mammal communities. Additionally, as ecologists grapple with the best metrics to measure communities, we expect that the diversity of small mammal communities will increasingly be measured in terms of functional and phylogenetic diversity. Finally, with a growing emphasis on the use of big data (Hampton et al. 2013) and remote sensing of ecological patterns (Stephenson 2019), we hope that there is still a place for the excellent experimental work on small mammals that has help shaped our understanding of important ecological processes.

REFERENCES

Abramsky, Z., M. I. Dyer, and P. D. Harrison. 1979. "Competition among small mammals in experimentally perturbed areas of the shortgrass prairie." *Ecology* 60 (3): 530–536. https://doi.org/10.2307/1936073.

Andersen, A. N., G. D. Cook, L. K. Corbett, M. M. Douglas, R. W. Eager, J. Russell-Smith, et al. 2005. "Fire frequency and biodiversity conservation in Australian tropical savannas: Implications from the Kapalga fire experiment." *Austral Ecology* 30 (2): 155–167. https://doi.org/10.1111/j.1442-9993.2005.01441.x.

Anderson, M. J., and D. C. I. Walsh. 2013. "PERMANOVA, ANOSIM, and the Mantel test in the face of heterogeneous dispersions: What null hypothesis are you testing?" *Ecological Monographs* 83 (4): 557–574. https://doi.org/10.1890/12-2010.1.

Anthony, R. G., L. J. Niles, and J. D. Spring. 1981. "Small-mammal associations in forested and old-field habitats: A quantitative comparison." *Ecology* 62 (4): 955–963. https://doi.org/10.2307/1936994.

Avenant, N. L., and P. Cavallini. 2007. "Correlating rodent community structure with ecological integrity, Tussen-die-Riviere Nature Reserve, Free State province, South Africa." *Integrative Zoology* 2 (4): 212–219. https://doi.org/10.1111/j.1749-4877.2007.00064.x.

Baselga, A. 2010. "Partitioning the turnover and nestedness components of beta diversity." *Global Ecology and Biogeography* 19 (1): 134–143. https://doi.org/10.1111/j.1466-8238.2009.00490.x.

Baselga, A., D. Orme, S. Villeger, J. De Bortoli, and F. Leprieur. 2017. "Partitioning beta diversity into turnover and nestedness components." *Package betapart*, Version: 1.4–1.

Beca, G., M. H. Vancine, C. S. Carvalho, F. Pedrosa, R. Souza C. Alves, et al. 2017. "High mammal species turnover in forest patches immersed in biofuel plantations." *Biological Conservation* 210: 352–359. https://doi.org/10.1016/j.biocon.2017.02.033.

Blois, J. L., J. L. McGuire, and E. A. Hadly. 2010. "Small mammal diversity loss in response to late-Pleistocene climatic change." *Nature* 465 (7299): 771-U5. https://doi.org/10.1038/nature09077.

Bovendorp, R. S., F. T. Brum, R. A. McCleery, B. Baiser, R. Loyola, M. V. Cianciaruso, and M. Galetti. 2019. "Defaunation and fragmentation erode small mammal diversity dimensions in tropical forests." *Ecography* 42 (1): 23–35. https://doi.org/10.1111/ecog.03504.

Bovendorp, R. S., R. A. McCleery, and M. Galetti. 2017. "Optimising sampling methods for small mammal communities in Neotropical rainforests." *Mammal Review* 47 (2): 148–158. https://doi.org/10.1111/mam.12088.

Briani, D. C., A. R. T. Palma, E. M. Vieira, and R. P. B. Henriques. 2004. "Post-fire succession of small mammals in the Cerrado of central Brazil." *Biodiversity and Conservation* 13 (5): 1023–1037. https://doi.org/10.1023/B:BIOC.0000014467.27138.0b.

Bricker, M., D. Pearson, and J. Maron. 2010. "Small-mammal seed predation limits the recruitment and abundance of two perennial grassland forbs." *Ecology* 91 (1): 85–92. https://doi.org/10.1890/08-1773.1.

Brown, J. H., and E. J. Heske. 1990. "Temporal changes in a Chihuahuan desert rodent community." *Oikos* 59 (3): 290–302. https://doi.org/10.2307/3545139.

Brown, J. H., and J. C. Munger. 1985. "Experimental manipulation of a desert rodent community: Food addition and species removal." *Ecology* 66 (5): 1545–1563. https://doi.org/10.2307/1938017.

Brown, J. H., T. G. Whitham, S. K. M. Ernest, and C. A. Gehring. 2001. "Complex species interactions and the dynamics of ecological systems: Long-term experiments." *Science* 293 (5530): 643–650. https://doi.org/10.1126/science.293.5530.643.

Brum, F. T., C. H. Graham, G. C. Costa, S. B. Hedges, C. Penone, V. C. Radeloff, et al. 2017. "Global priorities for conservation across multiple dimensions of mammalian diversity." *Proceedings of the National Academy of Sciences of the United States of America* 114 (29): 7641–7646. https://doi.org/10.1073/pnas.1706461114.

Burel, F., J. Baudry, A. Butet, P. Clergeau, Y. Delettre, D. Le Coeur, et al. 1998. "Comparative biodiversity along a gradient of agricultural landscapes." *Acta Oecologica* 19 (1): 47–60. https://doi.org/10.1016/s1146–609x(98)80007–6.

Burel, F., A. Butet, Y. R. Delettre, and N. M. de la Pena. 2004. "Differential response of selected taxa to landscape context and agricultural intensification." *Landscape and Urban Planning* 67 (1–4): 195–204. https://doi.org/10.1016/s0169–2046(03)00039-2.

Cavia, R., G. R. Cueto, and O. V. Suarez. 2009. "Changes in rodent communities according to the landscape structure in an urban ecosystem." *Landscape and Urban Planning* 90 (1–2): 11–19. https://doi.org/10.1016/j.landurbplan.2008.10.017.

Chao, A. 2005. "Species estimation and applications" In *Encyclopedia of Statistical Sciences*, edited by S. Kotz, N. Balakrishnan, C. B. Readand, and B. Vidakovic, 7907–7916. New York: Wiley.

Chao, A., R. L. Chazdon, R. K. Colwell, and T. J. Shen. 2005. "A new statistical approach for assessing similarity of species composition with incidence and abundance data." *Ecology Letters* 8: 148–159.

Chao, A., and C.-H. Chiu. 2016. "Species richness: Estimation and comparison." In *Wiley StatsRef: Statistics Reference Online*, edited by N. Balakrishnan, T. Colton, B. Everitt, W. Piegorsch, F. Ruggeri, and J. L. Teugels, 1–26. https://doi.org/10.1002/9781118445112.stat03432.pub2.

Chazdon, R. L., R. K. Colwell, J. S. Denslow, and M. R. Guariguata. 1998. "Statistical methods for estimating species richness of woody regeneration in primary and secondary rain forests of northeastern Costa Rica." In *Forest Biodiversity Research, Monitoring and Modeling. Conceptual Background and Old World Case Studies*, edited by F. Dallmeier and J. A. Comiskey, 285–309. Paris: Parthenon Publishing.

Churchfield, S., J. Hollier, and V. K. Brown. 1991. "The effects of small mammal predators on grassland invertebrates, investigated by field exclosure experiment." *Oikos* 60 (3): 283–290. https://doi.org/10.2307/3545069.

Clarke, K. R., R. N. Gorley, P. J. Somerfield, and R. M. Warwick. 2014. *Change in Marine Communities: An Approach to Statistical Analysis and Interpretation*. Albany, New Zealand: Primer-E Ltd.

Clotfelter, E. D., A. B. Pedersen, J. A. Cranford, N. Ram, E. A. Snajdr, V. Nolan Jr., and E. D. Ketterson. 2007. "Acorn mast drives long-term dynamics of rodent and songbird populations." *Oecologia* 154 (3): 493–503. https://doi.org/10.1007/s00442-007-0859-z.

Crowner, A. W., and G. W. Barrett. 1979. "Effects of fire on the small mammal component of an experimental grassland community." *Journal of Mammalogy* 60 (4): 803–813. https://doi.org/10.2307/1380195.

Dambros, C. S., N. C. Caceres, L. Magnus, and N. J. Gotelli. 2015. "Effects of neutrality, geometric constraints, climate, and habitat quality on species richness and composition of Atlantic Forest small-mammals." *Global Ecology and Biogeography* 24 (9): 1084–1093. https://doi.org/10.1111/geb.12330.

Darracq, A. K., L. M. Conner, J. S. Brown, and R. A. McCleery. 2016. "Cotton rats alter foraging in response to an invasive ant." *PloS ONE* 11 (9): e0163220.

de la Pena, N. M., A. Butet, Y. Delettre, G. Paillat, P. Morant, L. Le Du, and F. Burel. 2003. "Response of the small mammal community to changes in western French agricultural landscapes." *Landscape Ecology* 18 (3): 265–278. https://doi.org/10.1023/a:1024452930326.

de la Sancha, N. U. 2014. "Patterns of small mammal diversity in fragments of subtropical Interior Atlantic Forest in eastern Paraguay." *Mammalia* 78 (4): 437–449. https://doi.org/10.1515/mammalia-2013-0100.

Dreiss, L. M., K. R. Burgio, L. M. Cisneros, B. T. Klingbeil, B. D. Patterson, S. J. Presley, and M. R. Willig. 2015. "Taxonomic, functional, and phylogenetic dimensions of rodent biodiversity along an extensive tropical elevational gradient." *Ecography* 38 (9): 876–888. https://doi.org/10.1111/ecog.00971.

Dueser, R. D., and H. H. Shugart Jr. 1978. "Microhabitats in a forest-floor small mammal fauna." *Ecology* 59 (1): 89–98. https://doi.org/10.2307/1936634.

Eccard, J. A., and H. Ylonen. 2003. "Interspecific competition in small rodents: From populations to individuals." *Evolutionary Ecology* 17 (4): 423–440. https://doi.org/10.1023/a:1027305410005.

Ernest, S. K. M., J. H. Brown, and R. R. Parmenter. 2000. "Rodents, plants, and precipitation: Spatial and temporal dynamics of consumers and resources." *Oikos* 88 (3): 470–482. https://doi.org/10.1034/j.1600-0706.2000.880302.x.

Ernest, S. K. M., G. M. Yenni, G. Allington, E. M. Christensen, K. Geluso, J. R. Goheen, et al. 2016. "Long-term monitoring and experimental manipulation of a Chihuahuan desert ecosystem near Portal, Arizona (1977–2013)." *Ecology* 97 (4): 1082–1082. https://doi.org/10.1890/15-2115.1.

Faller, C. R., and R. A. McCleery. 2017. "Urban land cover decreases the occurrence of a wetland endemic mammal and its associated vegetation." *Urban Ecosystems* 20 (3): 573–580. https://doi.org/10.1007/s11252-016-0626<H?1.

Fischer, C., C. Thies, and T. Tscharntke. 2011. "Small mammals in agricultural landscapes: Opposing responses to farming practices and landscape complexity." *Biological Conservation* 144 (3): 1130–1136. https://doi.org/10.1016/j.biocon.2010.12.032.

Fitzgibbon, C. D. 1997. "Small mammals in farm woodlands: The effects of habitat, isolation and surrounding land-use patterns." *Journal of Applied Ecology* 34 (2): 530–539. https://doi.org/10.2307/2404895.

Flowerdew, J. R., and S. A. Ellwood. 2001. "Impacts of woodland deer on small mammal ecology." *Forestry* 74 (3): 277–287. https://doi.org/10.1093/forestry/74.3.277.

Fox, B. J. 1990. "Changes in the structure of mammal communities over successional time scales." *Oikos* 59 (3): 321–329. https://doi.org/10.2307/3545142.

Garcia-Navas, V. 2019. "Phylogenetic and functional diversity of African muroid rodents at different spatial scales." *Organisms, Diversity and Evolution* 19: 637–650.

Gardener, M. 2014. *Community Ecology: Analytical Methods Using R and Excel*. Exeter: Pelagic Publishing.

Gitzen, R. A., S. D. West, C. C. Maguire, T. Manning, and C. B. Halpern. 2007. "Response of terrestrial small mammals to varying amounts and patterns of green-tree retention in Pacific Northwest forests." *Forest Ecology and Management* 251 (3): 142–155. https://doi.org/10.1016/j.foreco.2007.05.028.

Gomes, V., R. Ribeiro, and M. A. Carretero. 2011. "Effects of urban habitat fragmentation on common small mammals: Species versus communities." *Biodiversity and Conservation* 20 (14): 3577–3590. https://doi.org/10.1007/s10531-011-0149-2.

Goodman, S. M., and D. Rakotondravony. 2000. "The effects of forest fragmentation and isolation on insectivorous small mammals (Lipotyphla) on the Central High Plateau of Madagascar." *Journal of Zoology* 250: 193–200. https://doi.org/10.1017/s0952836900002041.

Gortat, T., M. Barkowska, A. Gryczynska-Siemiatkowska, A. Pieniazek, A. Kozakiewicz, and M. Kozakiewicz. 2014. "The effects of urbanization—small mammal communities in a gradient of human pressure in Warsaw city, Poland." *Polish Journal of Ecology* 62 (1): 163–172. https://doi.org/10.3161/104.062.0115.

Gotelli, N. J., and A. Chao. 2013. "Measuring and estimating species richness, species diversity, and biotic similarity from sampling data." In *The Encyclopedia of Biodiversity*, 2nd ed., vol. 5, edited by S. A. Levin, 195–211. Waltham, MA: Academic Press.

Gotelli, N. J., and R. K. Colwell. 2011. "Estimating species richness." In *Biological Diversity: Frontiers in Measurement and Assessment*, edited by E. Magurran and B. J. McGill, 39–54. New York: Oxford University Press.

Grant, W. E., N. R. French, and D. M. Swift. 1977. "Response of a small mammal community to water and nitrogen treatments in a shortgrass prairie ecosystem." *Journal of Mammalogy* 58 (4): 637–652. https://doi.org/10.2307/1380011.

Hampton, S. E., C. A. Strasser, J. J. Tewksbury, W. K. Gram, A. E. Budden, A. L. Batcheller, et al. 2013. "Big data and the future of ecology." *Frontiers in Ecology and the Environment* 11 (3): 156–162.

Hanski I., P. Tuchin, E. Korpimaki, and H. Henttonen. 1993. "Population oscillations of boreal rodents: Regulation by mustelid predators leads to chaos." *Nature* 364 (6434): 232–235.

Harris, D. B. 2009. "Review of negative effects of introduced rodents on small mammals on islands." *Biological Invasions* 11 (7): 1611–1630. https://doi.org/10.1007/s10530-008-9393-0.

Heaney, L. R. 2001. "Small mammal diversity along elevational gradients in the Philippines: An assessment of patterns and hypotheses." *Global Ecology and Biogeography* 10 (1): 15–39. https://doi.org/10.1046/j.1466-822x.2001.00227.x.

Heroldova, M., J. Bryja, J. Zejda, and E. Tkadlec. 2007. "Structure and diversity of small mammal communities in agriculture landscape." *Agriculture Ecosystems & Environment* 120 (2–4): 206–210. https://doi.org/10.1016/j.agee.2006.09.007.

Heske, E. J. 1995. "Mammalian abundances on forest-farm edges versus forest interiors in southern Illinois: Is there an edge effect?" *Journal of Mammalogy* 76: 562–568.

Hill, T. C. J., K. A. Walsh, J. A. Harris, and B. F. Moffett. 2003. "Using ecological diversity measures with bacterial communities." *FEMS Microbiology Ecology* 43 (1): 1–11. https://doi.org/10.1111/j.1574-6941.2003.tb01040.x.

Hortal, J., P. A. V. Borges, and C. Gaspar. 2006. "Evaluating the performance of species richness estimators: Sensitivity to sample grain size." *Journal of Animal Ecology* 75 (1): 274–287. https://doi.org/10.1111/j.1365-2656.2006.01048.x.

Hurst, Z. M., R. A. McCleery, B. A. Collier, R. J. Fletcher Jr., N. J. Silvy, P. J. Taylor, and A. Monadjem. 2013. "Dynamic edge effects in small mammal communities across a conservation-agricultural interface in Swaziland." *PLoS ONE* 8 (9): e74520. https://doi.org/10.1371/journal.pone.0074520.

Hurst, Z. M., R. A. McCleery, B. A. Collier, N. J. Silvy, P. J. Taylor, and A. Monadjem. 2014. "Linking changes in small mammal communities to ecosystem functions in an agricultural landscape." *Mammalian Biology* 79 (1): 17–23. https://doi.org/10.1016/j.mambio.2013.08.008.

Jost, L., A. Chao, and R. L. Chazdon. 2011. "Compositional similarity and beta (beta) diversity." In *Biological Diversity: Frontiers in Measurement and Assessment,* edited by A. Magürran and B. McGill, 66–84. New York: Oxford University Press.

Kalies, E. L., B. G. Dickson, C. L. Hambers, and W. W. Covington. 2012. "Community occupancy responses of small mammals to restoration treatments in ponderosa pine forests, northern Arizona, USA." *Ecological Applications* 22 (1): 204–217. https://doi.org/10.1890/11-0758.1.

Keesing, F. 1998. "Impacts of ungulates on the demography and diversity of small mammals in central Kenya." *Oecologia* 116 (3): 381–389.

Kéry, M., and J. A. Royle. 2015. *Applied Hierarchical Modeling in Ecology: Analysis of Distribution, Abundance and Species Richness in R and BUGS,* vol. 1: *Prelude and Static Models.* Waltham: Academic Press.

Kirkland, G. L., Jr. 1990. "Patterns of initial small mammal community change after clearcutting of temperate North American forests." *Oikos* 59 (3): 313–320.

Korpimaki, E., P. R. Brown, J. Jacob, and R. P. Pech. 2004. "The puzzles of population cycles and outbreaks of small mammals solved?" *Bioscience* 54 (12): 1071–1079. https://doi.org/10.1641/0006-3568(2004)054[1071:tpopca]2.0.co;2.

Korpimaki, E., and K. Norrdahl. 1998. "Experimental reduction of predators reverses the crash phase of small-rodent cycles." *Ecology* 79 (7): 2448–2455. https://doi.org/10.1890/0012-9658(1998)079[2448:eroprt]2.0.co;2.

Korpimaki, E., K. Norrdahl, O. Huitu, and T. Klemola. 2005. "Predator-induced synchrony in population oscillations of coexisting small mammal species." *Proceedings of the Royal Society B-Biological Sciences* 272 (1559): 193–202. https://doi.org/10.1098/rspb.2004.2860.

Krebs, C. J. 2014. *Ecological Methodology.* 3rd ed. Menlo Park, CA: Addison-Wesley Educational Publishers

Krebs, C. J., M. S. Gaines, B. L. Keller, J. H. Myers, and R. H. Tamarin. 1973. "Population cycles in small rodents." *Science* 179 (4068): 35–41. https://doi.org/10.1126/science.179.4068.35.

Krebs, C. J., B. L. Keller, and R. H. Tamarin. 1969. "*Microtus* population biology: Demographic changes in fluctuating populations of *M. ochrogaster* and *M. pennsylvanicus* in southern Indiana." *Ecology* 50 (4): 587–607. https://doi.org/10.2307/1936248.

Laliberté, E., and P. Legendre. 2010. "A distance-based framework for measuring functional diversity from multiple traits." *Ecology* 91 (1): 299–305. https://doi.org/10.1890/08-2244.1.

Legendre, P., and M. J. Anderson. 1999. "Distance-based redundancy analysis: Testing multispecies responses in multifactorial ecological experiments." *Ecological Monographs* 69 (1): 1–24. https://doi.org/10.1890/0012-9615(1999)069[0001:dbratm]2.0.co;2.

Leirs, H., R. Verhagen, W. Verheyen, P. Mwanjabe, and T. Msibe. 1996. "Forecasting rodent outbreaks in Africa: An ecological basis for *Mastomys* control in Tanzania." *Journal of Applied Ecology* 33 (5): 937–943.

Loggins, A. A., A. Monadjem, L. M. Kruger, B. E. Reichert, and R. A. McCleery. 2019. "Vegetation structure shapes small mammal communities in African savannas." *Journal of Mammalogy* 100 (4): 1243–1252.

Luza, A. L., G. L. Goncalves, and S. M. Hartz. 2015. "Phylogenetic and morphological relationships between nonvolant small mammals reveal assembly processes at different spatial scales." *Ecology and Evolution* 5 (4): 889–902. https://doi.org/10.1002/ece3.1407.

MacArthur, R. H. 1965. "Patterns of species diversity." *Biological Reviews* 40 (4): 510–533. https://doi.org/10.1111/j.1469–185X.1965.tb00815.x.

MacKenzie, D. I., J. D. Nichols, J. A. Royle, K. H. Pollock, L. Bailey, and J. E Hines. 2017. *Occupancy Estimation and Modeling: Inferring Patterns and Dynamics of Species Occurrence.* Waltham: Academic Press.

Magurran, A. E., and B. J. McGill. 2011. *Biological Diversity: Frontiers in Measurement and Assessment.* New York: Oxford University Press.

Maisonneuve, C., and S. Rioux. 2001. "Importance of riparian habitats for small mammal and herpetofaunal communities in agricultural landscapes of southern Québec." *Agriculture Ecosystems & Environment* 83 (1–2): 165–175. https://doi.org/10.1016/s0167-8809(00)00259-0.

Mamba, M., N. J. Fasel, A. M. Themb'alilahlwa, J. D. Austin, R. A. McCleery, and A. Monadjem. 2019. "Influence of sugarcane plantations on the population dynamics and community structure of small mammals in a savanna-agricultural landscape." *Global Ecology and Conservation* 20 (2019): e00752.

Mares, M. A., K. A. Ernest, and D. D. Gettinger. 1986. "Small mammal community structure and composition in the Cerrado province of central Brazil." *Journal of Tropical Ecology* 2 (4): 289–300.

Maurer, B. A., and B. J. McGill. 2011. "Measurement of species diversity." In *Biological Diversity: Frontiers in Measurement and Assessment*, edited by A. E. Magurran and B. J. McGill, 56–65. New York: Oxford University Press.

McArdle, B. H., and M. J. Anderson. 2001. "Fitting multivariate models to community data: A comment on distance-based redundancy analysis." *Ecology* 82 (1): 290–297. https://doi.org/10.1890/0012-9658(2001)082[0290:fmmtcd]2.0.co;2.

McCain, C. M. 2004. "The mid-domain effect applied to elevational gradients: Species richness of small mammals in Costa Rica." *Journal of Biogeography* 31 (1): 19–31. https://doi.org/10.1046/j.0305-0270.2003.00992.x.

McCleery, R. A., A. R. Holdorf, L. L. Hubbard, and B. D. Peer. 2015. "Maximizing the wildlife conservation value of road right-of-ways in an agriculturally dominated landscape." *PLoS ONE* 10 (3). https://doi.org/10.1371/journal.pone.0120375.

McCleery, R. A., A. Monadjem, B. Baiser, R. Fletcher Jr., K. Vickers, and L. Kruger. 2018. "Animal diversity declines with broad-scale homogenization of canopy cover in African savannas." *Biological Conservation* 226: 54–62. https://doi.org/10.1016/j.biocon.2018.07.020.

McCune, B., and J. B. Grace. 2002. *Analysis of Ecological Communities*. Gleneden Beach: Mjm Software Design.

McGarigal, K., S. Cushman, and S. Stafford. 2000. *Multivariate Statistics for Wildlife and Ecology Research*. New York: Springer.

McGregor, R. L., D. J. Bender, and L. Fahrig. 2008. "Do small mammals avoid roads because of the traffic?" *Journal of Applied Ecology* 45 (1): 117–123. https://doi.org/10.1111/j.1365-2664.2007.01403.x.

McNaughton, S. J., and L. L. Wolf. 1970. "Dominance and the niche in ecological systems." *Science* 167 (3915): 131–9.

Meserve, P. L., W. B. Milstead, and J. R. Gutierrez. 2001. "Results of a food addition experiment in a north-central Chile small mammal assemblage: Evidence for the role of 'bottom-up' factors." *Oikos* 94 (3): 548–556. https://doi.org/10.1034/j.1600-0706.2001.940316.x.

Michel, N., F. Burel, and A. Butet. 2006. "How does landscape use influence small mammal diversity, abundance and biomass in hedgerow networks of farming landscapes?" *Acta Oecologica* 30 (1): 11–20. https://doi.org/10.1016/j.actao.2005.12.006.

Michel, N., F. Burel, P. Legendre, and A. Butet. 2007. "Role of habitat and landscape in structuring small mammal assemblages in hedgerow networks of contrasted farming landscapes in Brittany, France." *Landscape Ecology* 22 (8): 1241–1253. https://doi.org/10.1007/s10980-007-9103-9.

Miller, D. A., R. E. Thill, M. A. Melchiors, T. B. Wigley, and P. A. Tappe. 2004. "Small mammal communities of streamside management zones in intensively managed pine forests of Arkansas." *Forest Ecology and Management* 203 (1–3): 381–393. https://doi.org/10.1016/j.foreco.2004.08.007.

Misra, M. K., and B. N. Misra. 1981. "Species diversity and dominance in a tropical grassland community." *Folia Geobotanica and Phytotaxonomica* 16 (3): 309–316.

Monadjem, A. 1999. "Geographic distribution patterns of small mammals in Swaziland in relation to abiotic factors and human land-use activity." *Biodiversity and Conservation* 8: 223–237.

Monadjem, A., J. Decher, W.-Y. Crawley, and R. A. McCleery. 2018. "The conservation status of a poorly known range-restricted mammal, the Nimba otter-shrew *Micropotamogale lamottei*." *Mammalia* 83 (1): 1–10.

Monadjem, A., and M. R. Perrin. 2003. "Population fluctuations and community structure of small mammals in a Swaziland grassland over a three-year period." *African Zoology* 38 (1): 127–137.

Moritz, C., J. L. Patton, C. J. Conroy, J. L. Parra, G. C. White, and S. R. Beissinger. 2008. "Impact of a century of climate change on small-mammal communities in Yosemite National Park, USA." *Science* 322 (5899): 261–264. https://doi.org/10.1126/science.1163428.

Morris, G., J. A. Hostetler, L. M. Conner, and M. K. Oli. 2011. "Effects of prescribed fire, supplemental feeding, and mammalian predator exclusion on hispid cotton rat populations." *Oecologia* 167 (4): 1005–1016. https://doi.org/10.1007/s00442-011-2053-6.

Morris, E. K., T. Caruso, F. Buscot, M. Fischer, C. Hancock, T. S. Maier, et al. 2014. "Choosing and using diversity indices: Insights for ecological applications from the German Biodiversity Exploratories." *Ecology and Evolution* 4 (18): 3514–3524.

Moses, R. A., and S. Boutin. 2001. "The influence of clear-cut logging and residual leave material on small mammal populations in aspen-dominated boreal mixed-

woods." *Canadian Journal of Forest Research* 31 (3): 483–495. https://doi.org/10.1139
/cjfr-31-3-483.

Myers, P., B. L. Lundrigan, S. M. G. Hoffman, A. P. Haraminac, and S. H. Seto. 2009.
"Climate-induced changes in the small mammal communities of the Northern
Great Lakes Region." *Global Change Biology* 15 (6): 1434–1454. https://doi.org/10.1111
/j.1365-2486.2009.01846.x.

Nor, S. M. 2001. "Elevational diversity patterns of small mammals on Mount Kinabalu,
Sabah, Malaysia." *Global Ecology and Biogeography* 10 (1): 41–62. https://doi.org
/10.1046/j.1466-822x.2001.00231.x.

Nupp, T. E., and R. K. Swihart. 2000. "Landscape-level correlates of small-mammal assemblages in forest fragments of farmland." *Journal of Mammalogy* 81 (2): 512–526.
https://doi.org/10.1644/1545-1542(2000)081<0512:llcosm>2.0.co;2.

Ostfeld, R. S., R. H. Manson, and C. D. Canham. 1997. "Effects of rodents on survival of
tree seeds and seedlings invading old fields." *Ecology* 78 (5): 1531–1542.

Ostoja, S. M., and E. W. Schupp. 2009. "Conversion of sagebrush shrublands to exotic
annual grasslands negatively impacts small mammal communities." *Diversity and
Distributions* 15 (5): 863–870. https://doi.org/10.1111/j.1472-4642.2009.00593.x.

Rickart, E. A., L. R. Heaney, D. S. Balete, and B. R. Tabaranza Jr. 2011. "Small mammal
diversity along an elevational gradient in northern Luzon, Philippines." *Mammalian Biology* 76 (1): 12–21. https://doi.org/10.1016/j.mambio.2010.01.006.

Risbey, D. A., M. C. Calver, J. Short, J. S. Bradley, and I. W. Wright. 2000. "The impact
of cats and foxes on the small vertebrate fauna of Heirisson Prong, Western Australia. II. A field experiment." *Wildlife Research* 27 (3): 223–235. https://doi.org/10.1071
/wr98092.

Rowe, K. C., K. M. C. Rowe, M. W. Tingley, M. S. Koo, J. L. Patton, C. J. Conroy, et al.
2015. "Spatially heterogeneous impact of climate change on small mammals of
montane California." *Proceedings of the Royal Society B-Biological Sciences* 282
(1799). https://doi.org/10.1098/rspb.2014.1857.

Rowe, R. J., R. C. Terry, and E. A. Rickart. 2011. "Environmental change and declining
resource availability for small-mammal communities in the Great Basin." *Ecology*
92 (6): 1366–1375. https://doi.org/10.1890/10-1634.1.

Saetnan, E. R., and C. Skarpe. 2006. "The effect of ungulate grazing on a small mammal
community in southeastern Botswana." *African Zoology* 41 (1): 9–16. https://doi.org
/10.1080/15627020.2006.11407331.

Sanchez-Giraldo, C., and J. F. Diaz-Nieto. 2015. "Dynamics of species composition of
small non-volant mammals from the northern Cordillera Central of Colombia."
Mammalia 79 (4): 385–397. https://doi.org/10.1515/mammalia-2014–0018.

Santos-Filho, M., C. A. Peres, D. J. da Silva, and T. M. Sanaiotti. 2012. "Habitat patch
and matrix effects on small-mammal persistence in Amazonian forest fragments."
Biodiversity and Conservation 21 (4): 1127–1147. https://doi.org/10.1007/s10531-012
-0248-8.

Schnurr, J. L., R. S. Ostfeld, and C. D. Canham. 2002. "Direct and indirect effects of
masting on rodent populations and tree seed survival." *Oikos* 96 (3): 402–410.
https://doi.org/10.1034/j.1600-0706.2002.960302.x.

Shure, D. J. 1971. "Tidal flooding dynamics: Its influence on small mammals in barrier beach marshes." *American Midland Naturalist* 85 (1): 36–44. https://doi.org
/10.2307/2423909.

Simelane, F. N., T. A. M. Mahlaba, J. T. Shapiro, D. MacFadyen, and A. Monadjem. 2018. "Habitat associations of small mammals in the foothills of the Drakensberg Mountains, South Africa." *Mammalia* 82 (2): 144–152. https://doi.org/10.1515/mammalia-2016-0130.

Smith, B., and J. B. Wilson. 1996. "A consumer's guide to evenness indices." *Oikos* 76 (1):70–82. https://doi.org/10.2307/3545749.

Sovie, A., L. M. Conner, J. S. Brown, and R. A. McCleery. 2021. "Increasing woody cover facilitates competitive exclusion of a savanna specialist." *Biological Conservation.* https://doi.org/10.1016/j.biocon.2021.108971.

Start, A. N., A. A. Burbidge, M. C. McDowell, and N. L. McKenzie. 2012. "The status of non-volant mammals along a rainfall gradient in the south-west Kimberley, Western Australia." *Australian Mammalogy* 34 (1): 36–48. https://doi.org/10.1071/am10026.

Stephens, R. B., D. J. Hocking, M. Yamasaki, and R. J. Rowe. 2017. "Synchrony in small mammal community dynamics across a forested landscape." *Ecography* 40 (10): 1198–1209. https://doi.org/10.1111/ecog.02233.

Stephenson, P. J. 1993. "The small mammal fauna of Réserve Spéciale d'Analamazaotra, Madagascar: The effects of human disturbance on endemic species diversity." *Biodiversity and Conservation* 2 (6): 603–615. https://doi.org/10.1007/bf00051961.

Stephenson, P. J. 2019. "Integrating remote sensing into wildlife monitoring for conservation." *Environmental Conservation* 46 (3): 181–183.

Suzan, G., E. Marce, J. T. Giermakowski, J. N. Mills, G. Ceballos, R. S. Ostfeld, et al. 2009. "Experimental evidence for reduced rodent diversity causing increased hantavirus prevalence." *PLoS ONE* 4 (5). https://doi.org/10.1371/journal.pone.0005461.

Swihart, R. K., and N. A. Slade. 1990. "Long-term dynamics of an early successional small mammal community." *American Midland Naturalist* 123 (2): 372–382. https://doi.org/10.2307/2426565.

Torre, I., M. Diaz, J. Martinez-Padilla, R. Bonal, J. Vinuela, and J. A. Fargallo. 2007. "Cattle grazing, raptor abundance and small mammal communities in Mediterranean grasslands." *Basic and Applied Ecology* 8 (6): 565–575. https://doi.org/10.1016/j.baae.2006.09.016.

Vellend, M., W. K. Cornwell, K. Magnuson-Ford, and A. O. Mooers. 2011. "Measuring phylogenetic biodiversity." In *Biological Diversity: Frontiers in Measurement and Assessment*, edited by A. E. Magurran and B. J. McGill, 194–207. New York: Oxford University Press.

Walther, B. A., and J. L. Moore. 2005. "The concepts of bias, precision and accuracy, and their use in testing the performance of species richness estimators, with a literature review of estimator performance." *Ecography* 28 (6): 815–829. https://doi.org/10.1111/j.2005.0906-7590.04112.x.

Weiher, E. 2011. "A primer of trait and functional diversity." In *Biological Diversity: Frontiers in Measurement and Assessment*, edited by A. E. Magurran and B. J. McGill, 175–193. New York: Oxford University Press.

Whitford, W. G. 1976. "Temporal fluctuations in density and diversity of desert rodent populations." *Journal of Mammalogy* 57 (2): 351–369. https://doi.org/10.2307/1379694.

Whittaker, R. H. 1972. "Evolution and measurement of species diversity." *Taxon* 21 (2/3): 213–251.

Williams, S. E., H. Marsh, and J. Winter. 2002. "Spatial scale, species diversity, and habitat structure: Small mammals in Australian tropical rain forest." *Ecology* 83 (5): 1317–1329. https://doi.org/10.2307/3071946.

Wilman, H., J. Belmaker, J. Simpson, C. de la Rosa, M. M. Rivadeneira, and W. Jetz. 2014. "EltonTraits 1.0: Species-level foraging attributes of the world's birds and mammals." *Ecology* 95 (7): 2027–2027.

11. Dietary Studies on Small Mammals

Background

Being able to determine the diet of an animal is critical to understanding many aspects of the animal's biology, tropic interactions, and ecological role in the community. Small mammals play a key role in many ecosystems, where they serve as the key link between the vegetative community and predators (Krebs et al. 2003). An understanding of small mammal diet is particularly important as it provides key insight into ecological and evolutionary factors, such as competitive interactions, habitat selection, and body condition (Krebs 1998). Traditionally, dietary studies of small mammals depended on either direct observation or examination of stomach contents and scat. While these direct methods are still used, indirect methods, particularly molecular genetics or isotope analysis, also provide opportunities for understanding foraging behavior and food sources of small mammals, many of which are highly secretive and difficult to study directly. The methods presented in this chapter are applicable across all the groups of small mammals covered by this book.

Aims

Here we review the various tools that are available for determining the diet and nutritional input of a small mammal. We start by exploring the strengths and weaknesses of traditional methods (such as direct observation and stomach/fecal analyses) and then transition to more recently developed methods, including stable isotopes and emerging molecular techniques. As with the rest of the book, the focus is on techniques for ecological research. Rats and mice have been used extensively in nutritional studies, particularly as models for human health; such studies are not dealt with in this chapter.

Direct Observation of Diet

Small mammals are, for the most part, not as well suited to direct observation as large mammals; however, some species are sufficiently large or conspicuous enough to observe as they forage and feed. The diets of a few such species of small mammals have been documented in this manner (de Villiers et al. 1994, Kuo and Lee 2003, Morrant and Petit 2012). Squirrels, pikas, and some other diurnal species that live in relatively open environments can be studied by simply watching them with binoculars or a spotting scope; however, determining the diets of most small mammals requires other approaches. For example, researchers walked tame Cape porcupines (*Hystrix africaeaustralis*) on a leash and recorded what they ate (de Villiers et al. 1994). In contrast, researchers used radiotelemetry to observe the Indian giant flying squirrel (*Petaurista philippensis*) and the western pygmy possum (*Cercartetus concinnus*) feeding (see Chapter 8) (Kuo and Lee 2003, Morrant and Petit 2012). Still, even if animals are observable, watching them forage is both time and labor intensive, and there is the potential that diet analyses may be biased toward consumption of larger items and that smaller food items may be missed, misidentified, or grouped into broad categories (Brown and Ewins 1996, Gonzalez-Solis et al. 1997).

Advantages to direct observations are that little equipment is required, and because observations are made directly, there is no need for sample processing. However, adequate training is needed to be able to identify plant (or other food) species from a distance. Adequate training is similarly required to identify functional groups (e.g., forbs, grasses), but they may be easier to classify than species. Regardless, direct observation of small mammal diet has been limited due to the cryptic nature of many small mammals.

Microhistology

Microhistology has been a primary means of studying the diets of mammals for decades (Baumgartner and Martin 1939). The approach relies on the researcher identifying plants species by their cell shapes or arrangements, or similarly from identifying arthropod and other prey items based on reference collections from known species. It is also possible to determine what an animal has eaten by examining its stomach contents. This has been the traditional approach used for mammals in general and small mammals in particular (Vaughan et al. 2011). This approach is a destructive method and often requires many animals to be killed. This is often possible in situations where researchers are studying pests in agricultural landscapes, studying diseases, or using road killed animals. After sacrificing the animal and removing its stomach, researchers clean the stomach contents (Chapter 4), stain them with a dye, and then usually examine them under a microscope. The various methods for preparing stomach contents for analysis are discussed in detail by Hansson (1970); anyone interested in conducted such work is urged to first read this important and thorough paper. This typically involves washing and sieving the stomach contents, and then macerating (clearing by boiling or application of acid), staining, mounting on microscope slide, and preserving (Hansson 1970) them.

Although it is possible to identify some dietary remains to species level, particularly for granivorous small mammals, most researchers have tended to categorize stomach contents into several broad categories such as "seed," "insect," "fruit," "vegetative plant material," and so on (Kincaid and Cameron 1982, Montgomery and Montgomery 1990, Kerley 1992, Wirminghaus and Perrin 1992, Monadjem 1997). Analyses based on such broad categories are often sufficient for trophic studies of small mammal communities (Kerley 1992). Knowing the broad category, or even specific species, in the diet of a small mammal may be useful for certain types of ecological questions but not for others (Murray et al. 1999). Different species (and individuals) of seeds or insects may vary in essential nutrients such as nitrogen or fat content, and it is these latter measures that some researchers are after when they wish to determine the quality of the diet (Cepelka et al. 2013, Janova et al. 2016). For example, dietary fat and protein are known to stimulate (or at least be necessary for) reproduction in small mammals (Field 1975). However, except in a few situations, it is not possible to get a species-level resolution through stomach content. One exception is when a small mammal's prey base is restricted to just a few species. For example, in the Kerguelen archipelago, a subantarctic island, just a few species of arthropods were present, including one species of spider, one aphid, one flightless moth, three weevils, and five flies, as well as two species of earthworms (Le Roux et al. 2002). Here it was possible to identify the arthropod fragments and earthworm chaetae to species in stomach contents of house mice (*Mus musculus*), something that would not have been possible in a more diverse community.

Certain diets also lend themselves more readily to identification than others. For example, arthropod remains can often be easily identified to order and sometimes family. Additionally, grass seeds and fruits can often be identified to species level, as long as a reference collection is available (Hansson

1970, Field 1975). However, in areas with high or poorly documented diversity, such as tropical rainforests and savannas, species-level identification is usually not possible or plausible through stomach-content analysis.

One way to examine stomach contents without sacrificing the animal is to conduct a stomach pump (Kronfeld and Dayan 1998). This method requires anesthetizing the animal and injecting saline into the animal's stomach, after which liquid contents can be removed. Perhaps because of the difficulties of setting up such a system, this method has not been frequently used. However, it is probably worth considering when needing to investigate the diet of endangered species, or species with small, fragile populations from which individuals cannot readily be harvested for destructive stomach-content analysis.

Mammals, as a result of their high metabolic rates, produce large quantities of feces, which have been used to determine diets of many species (Putman 1984). For small mammals, feces are often collected from the trap on first capture of an individual. Subsequent nights are usually disregarded because there is a considerable amount of bait in the fecal matter. One potential way to reduce the amount of bait in feces is to put out paper plates in appropriate habitats with a small amount of bait added to them. When rodents eat the bait, they defecate on the plate, but they then leave, so the bait is not defecated onto the plate. We have had success using this method of collecting feces from both hispid cotton rats (*Sigmodon hispidus*) and African savanna rodent communities (e.g., multimammate mice, *Mastomys natalensis*, and pygmy mice, *Mus minutoides*) (see Fig. 2.5). After collection, feces need to be softened, fixed (Vieira 2003), washed through sieves, and stained before being prepared into slides. This method is detailed in Carron et al. (1990), which should be referred to by anyone wishing to conduct such a study. A reference collection of potential food items is also useful for identification.

This technique has been well developed for a variety of larger mammals, where the species producing the fecal fragments are easier to identify due to their size. Fecal analysis of diet through traditional (microscopic) methods on small mammals has been limited. The diets of some murid rodent and cricetid species have been investigated with fecal analysis in Australia, North and South America, and Africa (Kincaid and Cameron 1982, Carron et al. 1990, Oguge 1995, Vieira 2003). Additionally, fecal analysis has been used to analyze of the diet common shrews (*Sorex araneus*) (Churchfield 1982); fecal pellets were collected from live-captured shrews and dietary fragments compared with a reference collection of invertebrates.

In studies on rodent diets, the fragments are categorized into broad groups instead of species, although the categories have been at a relatively fine resolution. For example, one study included categories for bark, monocotyledon leaves, dicotyledon leaves, and fungi (Carron et al. 1990). Similar to stomach-content analysis, fecal analysis may be useful for gaining a basic understand of the diet of small mammals; however, proportional differences in the use of different food are likely biased by differences in digestibility (Stewart 1967). For example, arthropods are often overestimated in fecal analysis because their cuticles are less likely to be digested than other materials. However, it still may be possible to identify differences between species and then adjust data by a coefficient of digestibility to improve results (Kincaid and Cameron 1982, Vieira 2003).

There are also differences that may arise related to temporal scale. Stomach-content analysis will provide a window on what has been consumed in the previous few hours, whereas fecal analysis may cover a longer period of time, perhaps a day. However, it is also possible to get insights into even longer periods, such as with stable isotopes (Davis and Pineda-Munoz 2016) (see below). The methods that we have described this far are what can be termed "traditional" methods. Rapid technological advances have made a whole suite of new methods available, which we discuss in the next few sections.

Stable Isotope Analysis

Based on an examination of the stomach contents of a rodent or shrew, we can make statements such as "a species is predominantly granivorous but increasing the intake of insects during the breeding season." However, if we wish to understand the interactions that take place between ecologically closely related species (e.g., testing ideas about resource partitioning), we may need a finer resolution of the diet. One way in which this has been achieved has been through the application of stable isotope analysis (SIA) (Parnell et al. 2010, Jackson et al. 2011), a technique that has been greatly refined for ecological studies since its first applications in the late 1970s (DeNiro and Epstein 1977, Kelly 2000). For example, recent studies have investigated competitive interactions and resource partitioning between rodent species by using SIA (Robb et al. 2012, Bauduin et al. 2013, Symes et al. 2013).

SIA measures the relative contributions of different nonradioactive isotopes in a particular tissue, such as bone, skin, fur, muscle, or organ (Drever et al. 2000, McKechnie 2004, Crawford et al. 2008, Shiels et al. 2013, Robb et al. 2016). The most commonly used isotopes in small mammal studies are those of the elements carbon (C) and nitrogen (N), although the stable isotopes of at least another five elements have been used in ecological studies, including oxygen, calcium, and sulfur (Ben-David and Flaherty 2012). Specifically, it is the ratio of C^{13} to C^{12} and N^{15} to N^{14} that we are interested in, which are designated as $\delta^{13}C$ and $\delta^{15}N$, respectively (Bearhop et al. 2004, Newsome et al. 2007), and the units of measurement are in ‰ (parts per thousand).

What makes SIA useful is the fact that the ratio of stable isotopes in the protein of an animal reflects the ratio of those same stable isotopes in the protein of its diet. Therefore, examining the ratio of stable isotopes in the tissues of a small mammal tells us about its diet, without a need to examine its stomach contents or feces. Furthermore, this technique allows us to sample tissues in a nondestructive and even noninvasive manner by, for example, clipping a little bit of fur or a claw (Robb et al. 2012, Dammhahn et al. 2013).

Stable isotopes can therefore provide dietary information on the material assimilated into the tissues of a small mammal (McKechnie 2004). Differentiation between different assimilated materials is made possible by the fact that different potential food items have different isotopic signatures (ratios). More enriched (higher) $\delta^{15}N$ values typically indicate a higher trophic position (i.e., carnivorous as opposed to herbivorous), whereas more enriched $\delta^{13}C$ values indicate a greater proportion of basal resources having come from C_4 plants (i.e., from grasslands as opposed to woodlands/forests). Hence, by plotting $\delta^{15}N$ (traditionally on the y-axis) against $\delta^{13}C$ values for each individual or species, one can view the niche breadth of the community, or how different components (individuals, populations, or species) compare in dietary terms (Fig. 11.1) (Ben-David and Flaherty 2012). Trophic relationships can be estimated by examining the $\delta^{15}N$ values, where on average the discrimination between different trophic levels amounts to a 3.4‰ increase (Ben-David and Flaherty 2012).

A shortcoming of SIA is that it cannot inform researchers what was ingested by a small mammal—only what was assimilated into the body of the animal (and more specifically into the tissue that was sampled for analysis). Hence, isotopic ratios in different tissues (that take different lengths of time to grow) may provide information on different periods in the life of the animal. Isotopic signatures from blood may represent foods that were recently eaten, whereas signatures from bone and fur may represent what was eaten weeks or months earlier (Bearhop et al. 2004). Nonetheless, for researchers interested in understanding an animal's diet, there have been efforts to reconstruct diets based on SIA, which involves the use of mixing models (Boecklen et al. 2011, Phillips 2012, Phillips et al. 2014, Stock et al. 2018), where isotopic ratios are converted into estimates of the food source contributions. This requires samples of potential food items in the environment to be sent for isotopic analysis along with tissues of the small mammal whose diet is in question. However, in many ecological studies, the actual diet of the small mammal is less important than other considerations, such as trophic level, niche breadth, or some other aspect that can be determined without the need for mixing models.

For example, SIA has revolutionized certain aspects of ecology, particularly testing hypotheses about niche theory. The Hutchinsonian niche has

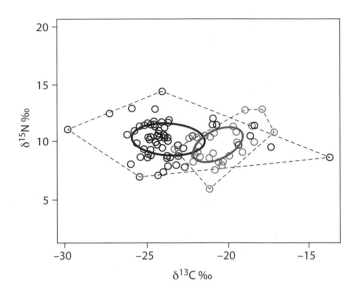

Fig. 11.1. Isotopic signatures of forest (black circles) and savanna (in gray circles) species showing separation in $\delta^{13}C$ but not in $\delta^{15}N$ (i.e., trophic levels)

been defined as an *n*-dimensional hypervolume where each dimension refers to some measurable component of the environment (Pulliam 2000). These dimensions have further been divided into environmental axes (such as rainfall and temperature) and bionomic axes (such as diet). It has been argued that the latter can be determined by stable isotopes (Bearhop et al. 2004, Newsome et al. 2007).

The analysis of dietary niche via SIA is now typically conducted using several R packages (Jackson et al. 2011). In particular, SIBER (stable isotope Bayesian ellipses in R) has become popular for calculating niche breadth (Jackson et al. 2011). The idea is to base niche breadth on isotopic signatures of a number of different individuals from the same population by calculating the spread of the points on the biplot (typically $\delta^{15}N$ against $\delta^{13}C$). Hence, in a community of small mammals, it is possible to identify generalists (with a large niche breadth) versus specialists (narrow niche breadth) (Dammhahn et al. 2017). There are in fact six metrics typically used to estimate community relationships from SIA (Newsome et al. 2007): (1) range in $\delta^{13}C$ values (providing information on the diversity of basal resources consumed); (2) range in $\delta^{15}N$ values (providing information on trophic structure); (3) total area of convex hull (niche breadth); (4) mean distance to centroid (species packing in a community); (5) mean nearest neighbor distance (density and clustering of species within a community); and (6) standard deviation of nearest neighbor distance (evenness of species packing and clustering) (Layman et al. 2007). In addition, isotopic diet analysis can provide information on seasonal or habitat-specific dietary shifts, or long-term dietary signals (Pompanon et al. 2012). These metrics have been used widely in research studies (Robb et al. 2012, 2016, Dammhahn et al. 2013). Convex hulls, however, are sensitive to differences in sample size. To address this bias, we recommend the use of Bayesian ellipses (Jackson et al. 2011). These ellipses are unbiased with respect to sample size and can easily be calculated in the R package SIBER.

In addition to the traditional niche metrics outlined above, it is possible to calculate niche overlap using a recently developed method (Cucherousset and Villéger 2015). To address the multitude of recently developed techniques, another new R package, mixSIAR, has recently been released that has unified numerous previous packages for SIA, including SIBER (Stock et al. 2018). Furthermore, it is important to note that SIA is not a magic formula that can convert any dataset into a cutting-edge manuscript on trophic interrelations within a community. As obvious as this might be, basic science still applies, and researchers should ensure that they have collected their data in the correct fashion for the SIA that they intend to conduct (Phillips et al. 2014). Additionally, isotopic signals do not reflect recent diets. Similarly, the signatures of some food items are weak and thus may not provide clear resolution of trophic linkages. In these cases, isotopic studies could be refined by complementary DNA sequencing methods to determine what was recently eaten.

Near Infrared Reflectance Spectroscopy

Near infrared reflectance spectroscopy (NIRS) can be used to identify plant forage quality (e.g., nitrogen intake) from ingesta or feces by measuring the chemical structure of the forage species. This approach has been used widely in studies of large ruminants, particularly for determining diet quality (Norris et al. 1976). There has been limited application on small mammals, in part due to the animals' small size and the need for relatively large samples for NIRS processing (Cepelka et al. 2017). However, it has recently been adapted and applied to small mammals (Cepelka et al. 2013, 2017, Janova et al. 2016).

Samples need to be homogenized prior to analysis in a spectrometer (e.g., Nicolet Antaris II FT-NIR analyzer) and a near infrared calibration and continuous validation is required in order to estimate nitrogen content (Dixon and Coates 2009), an aspect that Garnick et al. (2018) argue has been underappreciated in NIRS studies. However, the use of fecal samples allows for this approach to be applied to long-term monitoring of diet changes (relative nitrogen, fiber content, etc.). The approach works best on low diversity grazing or browsing diets, allowing for adequate calibrations to be developed and maintained (Garnick et al. 2018).

DNA Methods for Diet Determination

Advancements in DNA analysis over the past decade have greatly improved our ability to understand the diets and trophic linkages of small mammals. DNA barcoding is a term that refers to a now commonplace method of species identification using a standardized portion of DNA. The genes used have been demonstrated to be highly variable, providing the necessary resolution to distinguish among species. Furthermore, DNA primers can be readily developed to allow for either "universal" or group-specific amplification across a broad range of taxa (see Table 11.1 for dietary examples). The most common mtDNA gene region is the cytochrome c oxidase subunit 1 (COI) (Hebert et al. 2003). Plant barcoding has been less standardized (CBOL Plant Working Group1 2009, Soininen et al. 2009, Hollingsworth et al. 2011), at least until recently, with a few universal contenders becoming established. Currently, standard plant universal barcodes include *trnL*, *rbcL*, and *matK*, three chloroplast genes that have been utilized as universal polymerase chain reaction (PCR) primers to study plants in omnivore and herbivore mammals (De Barba et al. 2014).

The application of DNA barcoding approaches to gut contents and fecal samples has provided increased accuracy of dietary analysis but has the downside of labor-intensive PCR cloning and sequencing. Recently, "next-generation sequencing" (Chapter 15) has advanced genetic dietary analysis of small mammal diets by permitting the simultaneous identification of multiple dietary species without the need for time-consuming cloning and relatively low-throughput Sanger sequencing.

More typically, dietary breadth is of interest and numerous approaches have been applied to study diet, such as isotopic analysis (above), protein electrophoresis in carnivores, or alkane fingerprinting for herbivores (Pompanon et al. 2012). However, these methods have rarely been applied to small mammals and have largely been supplanted by advancements in DNA sequencing. The application of DNA technologies can address many study questions, ranging from targeted qualitative studies to broadly targeted quantitative metabarcoding projects. Pompanon et al. (2012) provide an overview of classifications of dietary studies utilizing DNA and summarize approaches into six categories, ranging from identifying individual food remains using targeted PCR to unsorted diet samples that are quantified via metabarcoding.

To identify prey species in a small mammal diet using a barcoding approach, the prey species will need to have been sequenced for the target gene beforehand, preferably from properly identified and vouchered specimens (Collins and Cruickshank 2013). Thus, for studies in poorly characterized regions, barcode databases will need to be developed before detailed dietary

Table 11.1. Barcoding information used to study mammal diet, including information on genome target (i.e., chloroplast DNA, cpDNA; nuclear genome, nDNA; mitochondrial, mtDNA) and universal (e.g., amplify broadly) or targeted (e.g., for resolving finer-scale species identity) resolution. All primer sequences and applications have been used to study the diet of mammals; applications involving small mammals are bold.

Taxa	Gene	Genome resolution	Primer sequence (5'-3')	Original source of primer	Mammal applications
Plants	*trnL*	Universal cpDNA	c—CGAAATCGGTAGACGCTACG d—GGGGATAGAGGGACTTGAAC	Taberlet et al. 1991	De Barba et al. 2014
	trnL	Universal cpDNA	g—GGGCAATCCTGAGCCAA h—CCATTGAGTCTCTGCACCTATC	Taberlet et al. 2007	**Soininen et al. 2009, 2013, 2017**; De Barba et al. 2014; Kartzinel et al. 2015; **Ozaki et al. 2018**; **Sato et al. 2018, 2019**
	trnH-psbA		psbAF—GTTATGCATGAACGTAATGCTC trnH2—CGCATGGTGGATTCACAATCC	Sang et al. 1997 Tate and Simpson 2003	**Schneider et al. 2017**
	rbcL	Universal cpDNA	rbcLa-F—ATGTCACCACAAACAGAGACTAAAGC rbcLa-R—GTAAAATCAAGTCCACCRCG	Levin 2003 Kress and Erickson 2007	**Khanam et al. 2016**
	rbcL	Universal cpDNA	rbcL h1aF—GGCAGCATTCCGAGTAACTCCTC rbcL h2aR—CGTCCTTTGTAACGATCAAG	Poinar et al. 1998	**Khanam et al. 2016**
	rbcL	Targeted cpDNA	See Dell'Agnello et al. 2019 *supplemental data for reference**	**Dell'Agnello et al. 2019**	**Dell'Agnello et al. 2019**
(Asteraceae)	ITS1	Targeted nDNA	ITS1-F—GATATCCGTTGCCGAGTC ITS1Ast-R—CGGCACGGCATGTGCCAAGG	Baamrane et al. 2012	Baamrane et al. 2012; De Barba et al. 2014
(Poaceae)	ITS1	Targeted nDNA	ITS1-F—GATATCCGTTGCCGAGTC ITS1Poa-R—CCGAAGGCGTCAAGGAACAC	Baamrane et al. 2012	Baamrane et al. 2012; De Barba et al. 2014
(Rosaceae)	ITS2	Targeted nDNA	ITS2Ros-F—YCTGCCTGGGCGTCACA ITS2Ros-R—CGTKVGYCGCCGAGGAC	De Barba et al. 2014	De Barba et al. 2014

Taxon	Marker	Type	Primers and sequences	Reference	Reference
(Cyperaceae)	ITS1	Targeted nDNA	ITS1-F—GATATCCGTTGCCGAGAGTC; ITS1Cyp-R—GGATGACGCCAAGGAACAC	Baamrane et al. 2012; De Barba et al. 2014	De Barba et al. 2014
	ITS2	Targeted nDNA	ITS1-F-rc—GACTCTCGGCAACGGATATC; ITS 4—TCCTCCGCTTATTGATATGC	Baamrane et al. 2012; White et al. 1990	De Barba et al. 2014
Invertebrates	16S	Universal mtDNA	16SMAV-F—CCAACATCGAGGTCRYAA; 16SMAV-R—ARTTACYNTAGGGATAACAG	De Barba et al. 2014	De Barba et al. 2014; Ozaki et al. 2018
		Mammal blocking primer	MamMAVB1—CCTAGGGATAACAGGCCAATCCTATT-C3		
	COI	Universal mtDNA	C1J1718-F—GGAGGATTTGGAAATTGATTAGTTCC; C1N2191-R—CCCCGGTAAAATTAAAATATAAACTTC; InvertCOI-F—GWATTTCHTCWATTYTWGGRG	Simon et al. 1994; Khanam et al. 2016	Khanam et al. 2016
	COI	Universal mtDNA	HCO—GGTCAACAAATCATAAAGATATTGG; LCO—TAAACTTCAGGGTGACCAAAAAATCA; Uni-MinibarF1—TCCACTAATCACAARGATATTGGTAC; UniMinibarR1—GAAAATCATAATGAAGGCATGAGC	Folmer et al. 1994; Meusnier et al. 2008	Hawlitschek et al. 2018
	COI	Universal mtDNA	ZBJ-ArtF1c—AGATATTGGAACWTTATATTTTATTTTTGG; ZBJ-ArtR2c—WACTAATCAATTWCCAAATCCTCC	Zeale et al. 2011	Sato et al. 2019
Vertebrates	12S	Universal mtDNA	12SV5—TTAGATACCCCACTATGC; 12SV5-R—TAGAACAGGCTCCTCTAG	Riaz et al. 2011	De Barba et al. 2014
		Human blocking primer	HomoB—CTATGCTTAGCCCTAAACCTCAACAGTTAAATCAACAAAACTGCT-C3	De Barba et al. 2014	
(Aves)	COX1	Universal mtDNA	BIRDF1—TTCTCCAACCACAAAGACATTGGCAC; AWCintR2—ATGTTGTTTATGAGTGGGAATGCTATG	Kerr et al. 2007; Patel et al. 2010	Khanam et al. 2016

analyses can be performed. Such a database would consist of DNA sequences from as many representative species as are deemed to be of importance. This can potentially be completed from existing herbarium collections, for example; however, field collections will be required for many taxa depending on the target species' known diet (e.g., whether they are insectivores, omnivores, or herbivores). A good source of DNA sequences is GenBank (https://www.ncbi.nlm.nih.gov/nucleotide/), where you can find sequences of almost any species that have been recently published. You can also use GenBank to "blast" (i.e., compare) the DNA sequences you have obtained from your study with previous studies. More valuable will be barcode-specific libraries such as that at the International Barcode of Life (https://ibol.org), which is a public collection of reference databases that is searchable for plants, insects, and other taxa.

Two general strategies for barcoding dietary analyses are using universal primers (e.g., that amplify all animal or all plant taxa) and using targeted group-specific primers (e.g., different primers designed for coleopterans, another for hymenopterans). Universal primers have clear advantages, including not requiring a priori knowledge about your focal species' dietary preferences. However, since general dietary preferences are likely to be known or readily deduced for most small mammals, either strategy should be feasible. For herbivorous small mammals, a universal plant barcoding approach has been developed that focuses on the P6 loop of the chloroplast *trn*L intron (Soininen et al. 2009). This method allows for substantially greater taxonomic resolution relative to microhistological approaches, for example, in the analysis of the diets of subarctic vole species (*Microtus oeconomus* and *Myodes rufocanus*) (Soininen et al. 2009) and in "pest" species like black rats (*Rattus rattus*) and house mice (*Mus musculus*) (Khanam et al. 2016).

A promising approach to studying diet composition is extended by metabarcoding and high-throughput sequencing (Table 11.2). This metabarcoding approach allows one to simultaneously identify dietary species from feces or stomach contents by directly sequencing millions of DNA strands using universal barcode primers. By pooling uniquely tagged PCR products (i.e., the amplifications from multiple fecal samples, or pooled PCR using different group-specific primers) and sequencing them together in a single sequencing run (Valentini et al. 2009), this type of metabarcoding approach makes processing of many samples simultaneously feasible and cost-effective. Metabarcoding has not been applied as widely to dietary studies of small mammals as for large mammals, though the method holds great promise for broader application. Recent examples include Ozaki et al. (2018), who used metabarcoding to evaluate the influence on dietary preference of wood mice (*Apodemus sylvaticus*) along a gradient of trace-metal contaminated soils. Compar-

Table 11.2. Aspects of DNA barcoding and metabarcoding relevant to dietary studies on small mammals. DNA barcoding refers to the application of taxon-specific primers to amplify an unidentified sample for the purpose of species identification. Metabarcoding refers to the joint amplification of pooled samples, either collected from environmental samples (e.g., water or soil) or from fecal or gut contents. The latter relies upon the high throughput of next-generation sequencing platforms (see Chapter 15).

	Typical application	Typical platform	Barcode identification
DNA Barcoding	Single or target species identification (e.g., to detect rare or weedy species in diet)	Sanger sequencing to produce ~600–800 bp of DNA sequence from target gene	Comparison of sequence to barcode database (e.g., BOLD Identification Engine)
Metabarcoding	Community reconstruction from multispecies DNA sample (e.g., soil, gut contents, feces)	Next-generation sequencing using universal taxon primers (e.g., Table 11.1)	Comparison of sequences to barcode database (e.g., BOLD, BLAST); tree-based identification (e.g., Hawlitschek et al. 2018)

ative metabarcoding studies on sympatric wood mice species (*A. argenteus* and *A. speciosus*) have revealed dietary niche partitioning (Sato et al. 2018). And, in a test to examine food-web dynamics, Schneider et al. (2017) found that native rodents preferred native plants over nonnative plants, despite the greater abundance of the latter, illustrating the foraging suppression of native plant diversity in a restored prairie. Interested readers should see Sato et al. (2019) for a recent discussion of the pros and cons of applying metabarcoding to study of small mammal diets.

Across mammal taxa (small and large), dietary analysis using DNA-based approaches continues to increase in popularity, largely because of the high dietary resolution over short time periods that is possible. Until recently, the ability to understand the relative proportion of items in a small mammal's diet remained limited to semi-quantitative measures such as frequencies of detection (see below). However, the potential for assessing dietary preference using quantitative PCR (qPCR), which monitors the abundance of amplified product during the PCR step (Deepak et al. 2007), has been demonstrated to reliably quantify dietary proportions (Thomas et al. 2016). The diet of Savi's pine voles (*Microtus savii*) was quantified using DNA barcoding combined with qPCR, and they were found to be selective feeders (Dell'Agnello et al. 2019). By adding qPCR as part of the genetic repertoire, more detailed information on dietary preferences and resource abundance can be inferred from small mammal diets (Dell'Agnello et al. 2019). For omnivorous or insectivorous small mammals, differentiating secondary predation remains problematic using DNA methods. For example, a diet of herbivorous insects eaten by omnivorous mammals will potentially bias dietary breadth.

Differentiating between What Is
Eaten and What Is Preferred

Diet studies based on observations of animals in the field can tell us what the animals eat and how much of it, but they do not inform selection of food items. To determine dietary selection, we need to know what foods were available to the animal and in what proportions (Rodgers 1990); such studies are predominantly conducted in the laboratory where the researcher can control food availability (which is extremely difficult in the field). These sorts of food-preference studies are typically called "cafeteria" studies and involve the researcher simultaneously placing a variety of foods of known (usually equal) quantity in a cage with a small mammal and recording how much of each food has been consumed after a set period (Vieira et al. 2015). A food selection index can then be calculated for each food item, which is then used to rank these foods from most to least selected (Perisse et al. 1989). An important review of the methodology for such food-choice experiments is given in Meier et al. (2012); although it was specifically written for ruminant livestock, the issues concerning design and methods are applicable to small mammals. Similar approaches have been used to study feeding preferences and consumption rates by voles (*Microtus arvalis* and *Myodes glareolus*) in agricultural settings (Fig. 11.2) (Fischer and Türke 2016).

The distinction between field-based dietary observation and laboratory-based choice experiments is important. What an animal eats in the field is not necessarily what it prefers to eat. There may be dozens of reasons for this discrepancy, including nonavailability of preferred food because of competitive interactions or fear of predation. By way of example, didelphid marsupials of the neotropics were traditionally treated as omnivores because their stomach contents typically contained many types of foods. However, based on food-choice experiments in the laboratory, the 12 didelphid species studied exhibited a range of preferences on a gradient from a tendency toward carnivory through to a tendency toward frugivory (Vieira et al. 2015). These differences, although subtle, may be important in understanding the community structure and trophic interactions of these small mammals. Of course, field-based observations and laboratory experiments are complementary, and both are needed for understanding small mammal ecology.

Conclusions and Future Directions

This chapter has summarized the techniques available for the examination and quantification of the diets of small mammals. Traditional studies relied on morphological observations, either direct behavioral observations of for-

Fig. 11.2. Seed-choice field experiment in an agricultural environment. Seeds are offered on a 10 cm × 10 cm seed depot divided into nine trays separated by wooden barriers to prevent seed mixing. Video placement next to a depot and screen capture of a vole at a seed depot (*inset*) shown. Experimental details are found in Fischer and Türke (2016).*(Photos: M. Türke / C. Fischer)*

aging or microhistological observations of stomach contents or feces, to determine what had been eaten. While relatively inexpensive, behavioral approaches are time-consuming and thus better suited for research questions rather than monitoring projects (Garnick et al. 2018). Microhistological work can also be labor intensive and requires a detailed reference collection for it

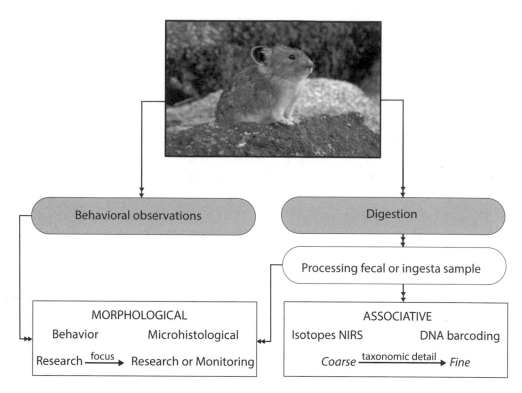

Fig. 11.3. To assess small mammal diets, morphological studies identify diet based on structural components of diet using behavioral observations (e.g., viewing and identifying forage) or the study of digesta (identification of plant cells or body parts). Because it is relatively difficult to observe the behavior of many small mammals, observation methods are best used for research on small mammal foraging behavior. Microhistological approaches are quantitative and can be processed fairly quickly, making them more suitable for longitudinal studies. For associative studies, techniques range from low taxonomic resolution (e.g., differentiating among grasses and forbs) to high resolution (identification of subspecies via DNA). *(Modified from Garnick et al. 2018; Pika photo: Diana Roberts)*

to be effective; however, it is the most quantitative of the all the methods involving sample processing (Fig. 11.3). Associative methods involving digested samples suffer from the issues connected with differential digestion and degradation of samples. NIRS requires complex equipment and careful development and maintenance of calibration sets to be considered reliable. However, it does permit rapid processing of large numbers of samples. Likewise, DNA barcoding (and metabarcoding) requires the establishment of an exhaustive reference database (when information on the total species diet composition is needed), requires access to complex equipment and resources, and usually provides qualitative information on diet (i.e., what species are present in the diet of an animal). Quantitative data is also possible but would require careful calibration, as different tissues have differential digestive rates and carry different amounts of DNA.

The ideal approach to studying small mammal diet will ultimately depend on the research priority and on the associated cost and sample sizes required to address the question. Coarse-scale data can be obtained readily using stable isotopes or NIRS approaches, obtaining information on functional groups, or studying grazing mammals whose diets are expected to consist of one or two components (Garnick et al. 2018). It remains a challenge to observe behavior for most small mammals due to their cryptic nature. Fine-scale quantitative data on diets (e.g., data to species level) can be obtained from microhistological approaches, with underrepresented dietary items identified from DNA barcoding (qualitatively). Although there is an increase in the popularity of DNA applications, the challenges associated with obtaining quantitative data and dealing with DNA degradation make barcoding a conservative approach (Garnick et al. 2018).

REFERENCES

Baamrane, M. A. A., W. Shehzad, A. Ouhammou, A. Abbad, M. Naimi, E. Coissac, et al. 2012. "Assessment of the food habits of the Moroccan dorcas gazelle in M'Sabih Talaa, west central Morocco, using the *trnL* approach." *PLoS ONE* 7 (4): e35643. https://doi.org/10.1371/journal.pone.0035643.

Bauduin, S., J. Cassaing, M. Issam, and C. Martin. 2013. "Interactions between the short-tailed mouse (*Mus spretus*) and the wood mouse (*Apodemus sylvaticus*): Diet overlap revealed by stable isotopes." *Canadian Journal of Zoology* 91: 102–109.

Baumgartner, L. L., and A. C. Martin. 1939. "Plant histology as an aid in squirrel food-habit studies." *Journal of Wildlife Management* 3: 266–268.

Bearhop, S., C. E. Adams, S. Waldron, R. A. Fuller, and H. Macleod. 2004. "Determining trophic niche width: A novel approach using stable isotope analysis." *Journal of Animal Ecology* 73: 1007–1012. https://doi.org/10.1111/j.0021-8790.2004.00861.x.

Ben-David, M., and E. A. Flaherty. 2012. "Stable isotopes in mammalian research: A beginner's guide." *Journal of Mammalogy* 93 (2): 312–328. https://doi.org/10.1644/11-MAMM-S-166.1.

Boecklen, W. J., C. T. Yarnes, B. A. Cook, and A. C. James. 2011. "On the use of stable isotopes in trophic ecology." *Annual Review of Ecology, Evolution, and Systematics* 42: 411–440. https://doi.org/10.1146/annurev-ecolsys-102209-144726.

Brown, K. M., and P. J. Ewins. 1996. "Technique-dependent biases in determination of diet composition: An example with ring-billed gulls." *Condor* 98: 34–41.

Carron, P. L., D. C. D. Happold, and T. M. Bubela. 1990. "Diet of two sympatric Australian subalpine rodents, *Mastacomys fuscus* and *Rattus fuscipes*." *Wildlife Research* 17: 479–489. https://doi.org/10.1071/WR9900479.

CBOL Plant Working Group1, P. M. Hollingsworth, L. L. Forrest, J. L. Spouge, M. Hajibabaei, S. Ratnasingham, et al. 2009. "A DNA barcode for land plants." *Proceedings of the National Academy of Sciences, USA* 106 (31): 12794–12797. https://doi.org/10.1073/pnas.0905845106.

Cepelka, L., M. Heroldová, E. Jánová, and J. Suchomel. 2013. "Dynamics of nitrogenous substance content in the diet of the wood mouse (*Apodemus sylvaticus*)." *Acta Universitatis Agriculturae et Silviculturae Mendelianae Brunensis* 61 (5): 1247–1253. https://doi.org/10.11118/actaun201361051247.

Cepelka, L., M. Heroldova, E. Janova, J. Suchomel, and D. Cizmar. 2017. "Rodent stomach sample preparation for nitrogen NIRS analysis." *Mammalian Biology* 87: 13–16.

Churchfield, S. 1982. "Food availability and the diet of the common shrew, *Sorex araneus*, in Britain." *Journal of Animal Ecology* 51 (1): 15–28.

Collins, R. A., and R. H. Cruickshank. 2013. "The seven deadly sins of DNA barcoding." *Molecular Ecology Resources* 13 (6): 969–975. https://doi.org/10.1111/1755-0998.12046.

Crawford, K., R. A. McDonald, and S. Bearhop. 2008. "Application of stable isotope techniques to the ecology of mammals." *Mammal Review* 38: 87–107.

Cucherousset, J., and S. Villéger. 2015. "Quantifying the multiple facets of isotopic diversity: New metrics for stable isotope ecology." *Ecological Indicators* 56: 152–160. https://doi.org/10.1016/j.ecolind.2015.03.032.

Dammhahn, M., T. M. Randriamoria, and S. M. Goodman. 2017. "Broad and flexible stable isotope niches in invasive non-native *Rattus* spp. in anthropogenic and natural habitats of Central Eastern Madagascar." *BMC Ecology* 17 (art. no. 16): 1–13. https://doi.org/10.1186/s12898-017-0125-0.

Dammhahn, M., V. Soarimalala, and S. M. Goodman. 2013. "Trophic niche differentiation and microhabitat utilization in a species-rich montane forest small mammal community of eastern Madagascar." *Biotropica* 45 (1): 111–118. https://doi.org/10.1111/j.1744-7429.2012.00893.x.

Davis, M., and S. Pineda-Munoz. 2016. "The temporal scale of diet and dietary proxies." *Ecology and Evolution* 6 (6): 1883–1897. https://doi.org/10.1002/ece3.2054.

De Barba, M., C. Miquel, F. Boyer, C. Mercier, D. Rioux, E. Coissac, et al. 2014. "DNA metabarcoding multiplexing and validation of data accuracy for diet assessment: Application to omnivorous diet." *Molecular Ecology Resources* 14 (2): 306–323. https://doi.org/10.1111/1755-0998.12188.

Deepak, S. A., K. R. Kottapalli, R. Rakwal, G. Oros, K. S. Rangappa, H. Iwahashi, et al. 2007. "Real-time PCR: Revolutionizing detection and expression analysis of genes." *Current Genomics* 8 (4): 234–251. https://doi.org/10.2174/138920207781386960.

Dell'Agnello, F., C. Natali, S. Bertolino, L. Fattorini, E. Fedele, B. Foggi, et al. 2019. "Assessment of seasonal variation of diet composition in rodents using DNA barcoding and real-time PCR." *Scientific Reports* 9 (art. no. 14124). https://doi.org/10.1038/s41598-019-50676-1.

DeNiro, M. J., and S. Epstein. 1977. "Mechanism of carbon isotope fractionation associated with lipid synthesis." *Science* 197: 261–263.

de Villiers, M. S., R. J. van Aarde, and H. M. Dott. 1994. "Habitat utilization by the Cape porcupine *Hystrix africaeaustralis* in a savanna ecosystem." *Journal of Zoology* 232: 539–549. https://doi.org/10.1111/j.1469-7998.1994.tb00002.x.

Dixon, R., and D. Coates. 2009. "Review: Near infrared spectroscopy of faeces to evaluate the nutrition and physiology of herbivores." *Journal of Near Infrared Spectroscopy* 17: 1–31.

Drever, M. C., L. K. Blight, K. A. Hobson, and D.F. Bertram. 2000. "Predation on seabird eggs by Keen's mice (*Peromyscus keeni*): Using stable isotopes to decipher the diet of a terrestrial omnivore on a remote offshore island." *Canadian Journal of Zoology* 78: 2010–2018. https://doi.org/10.1139/z00-131.

Field, A. C. 1975. "Seasonal changes in reproduction, diet and body composition of two equatorial rodents." *East African Wildlife Journal* 13: 221–235.

Fischer, C., and M. Türke. 2016. "Seed preferences by rodents in the agri-environment and implications for biological weed control." *Ecology and Environment* 6 (6): 5796–5807.

Folmer, O., M. Black, W. Hoeh, R. Lutz, and R. Vrijenhoek. 1994. "DNA primers for amplification of mitochondrial cytochrome c oxidase subunit I from diverse metazoan invertebrates." *Molecular Marine Biology and Biotechnology* 3: 294–299.

Garnick, S., P. S. Barboza, and J. W. Walker. 2018. "Assessment of animal-based methods used for estimating and monitoring rangeland herbivore diet composition." *Rangeland Ecology and Management* 71: 449–457.

Gonzalez-Solis, J., D. Oro, V. Pedrocchi, L. Jover, and X. Ruiz. 1997. "Bias associated with diet samples in Audouin's gulls." *Condor* 99: 773–779.

Hansson, L. 1970. "Methods of morphological diet micro-analysis in rodents." *Oikos* 21 (2): 255–266.

Hawlitschek, O., A. Fernández-González, A. Balmori-de la Puente, and J. Castresana. 2018. "A pipeline for metabarcoding and diet analysis from fecal samples developed for a small semi-aquatic mammal." *PLoS ONE* 13 (8): e0201763. https://doi.org/10.1371/journal.pone.0201763.

Hebert, P. D., A. Cywinska, S. L. Ball, and J. R. deWaard. 2003. "Biological identifications through DNA barcodes." *Proceedings of the Royal Society, Biological Sciences* 270 (1512): 313–321. https://doi.org/10.1098/rspb.2002.2218.

Holechek, J., M. Vavra, and R. D. Pieper. 1982. "Botanical composition determination of range herbivore diets: A review." *Journal of Range Management* 35: 309–315.

Hollingsworth, P. M., S. W. Graham, and D. P. Little. 2011. "Choosing and using a plant DNA barcode." *PLoS ONE* 6 (5): e19254. https://doi.org/10.1371/journal.pone.0019254.

Jackson, A. L., R. Inger, A. C. Parnell, and S. Bearhop. 2011. "Comparing isotopic niche widths among and within communities: SIBER—stable isotope Bayesian ellipses in R." *Journal of Animal Ecology* 80: 595–602. https://doi.org/10.1111/j.1365-2656.2011.01806.x.

Janova, E., M. Heroldova, and L. Cepelka. 2016. "Rodent food quality and its relation to crops and other environmental and population parameters in an agricultural landscape." *Science of the Total Environment* 562: 164–169. https://doi.org/10.1016/j.scitotenv.2016.03.165.

Kartzinel, T. R., P. A. Chen, T. C. Coverdale, D. L. Erickson, W. J. Kress, M. L. Kuzmina, et al. 2015. "DNA metabarcoding illuminates dietary niche partitioning by African large herbivores." *Proceedings of the National Academy of Sciences, USA* 112 (26): 8019–8024. https://doi.org/10.1073/pnas.1503283112.

Kelly, J. F. 2000. "Stable isotopes of carbon and nitrogen in the study of avian and mammalian trophic ecology." *Canadian Journal of Zoology* 78: 1–27. https://doi.org/10.1139/z99-165.

Kerley, G. I. H. 1992. "Trophic status of small mammals in the semi-arid Karoo, South Africa." *Journal of Zoology* 226 (4): 563–572. https://doi.org/10.1111/j.1469-7998.1992.tb07499.x.

Kerr, K. C. R., M. Y. Stoeckle, C. J. Dove, L. A. Weigt, C. M. Francis, and P. D. N. Hebert. 2007. "Comprehensive DNA barcode coverage of North American birds." *Molecular Ecology Notes* 7 (4): 535–543.

Khanam, S., R. Howitt, M. Mushtaq, and J. C. Russell. 2016. "Diet analysis of small mammal pests: A comparison of molecular and microhistological methods." *Integrative Zoology* 11: 98–110.

Kincaid, W. B., and G. N. Cameron. 1982. "Dietary variation in three sympatric rodents on the Texas coastal prairie." *Journal of Mammalogy* 63 (4): 668–672.

Krebs, C. 1998. "Niche measures and resource preferences." In *Ecological Methodology*, edited by E. Fogarty, V. McDougal, and N. Murray. San Francisco: Addison-Wesley.

Krebs, C., K. Danell, A. Angerbjorn, J. Agrell, D. Berteaux, K. A. Bråthen, et al. 2003. "Terrestrial trophic dynamics in the Canadian Arctic." *Canadian Journal of Zoology* 81: 827–843. https://doi.org/10.1139/z03-061.

Kress, W. J., and D. L. Erickson. 2007. "A two-locus global DNA barcode for land plants: The coding rbcL gene complements the non-coding trnH-psbA spacer region." *PLoS ONE* 6: e508.

Kronfeld, N., and T. Dayan. 1998. "A new method of determining diets of rodents." *Journal of Mammalogy* 79 (4): 1198–1202. https://doi.org/10.2307/1383011.

Kuo, C.-C., and L.-L. Lee. 2003. "Food availability and food habits of Indian giant flying squirrels (*Petaurista philippensis*) in Taiwan." *Journal of Mammalogy* 84 (4): 1330–1340. https://doi.org/10.1644/BOS-039.

Layman, C. A., D. A. Arrington, C. G. Montana, and D. M. Post. 2007. "Can stable isotope ratios provide for community-wide measures of trophic structure?" *Ecology* 88 (1): 42–48. https://doi.org/10.1890/0012-9658(2007)88[42:CSIRPF]2.0.CO;2.

Le Roux, V., J. L. Chapuis, Y. Frenot, and P. Vernon. 2002. "Diet of the house mouse (*Mus musculus*) on Guillou Island, Kerguelen Archipelago, Subantarctic." *Polar Biology* 25: 49–57. https://doi.org/10.1007/s003000100310.

Levin, R. A., W. L. Wagner, P. C. Hoch, M. Nepokroeff, J. C. Pires, E. A. Zimmer, et al. 2003. "Family-level relationships of Onagraceae based on chloroplast *rbcL* and *ndhF* data." *American Journal of Botany* 90 (1): 107–115. https://doi.org/10.3732/ajb.90.1.107.

McKechnie, A. E. 2004. "Stable isotopes: Powerful new tools for ecologists." *South African Journal of Science* 100: 131–134.

Meier, J. S., M. Kreuzer, and S. Marquardt. 2012. "Design and methodology of choice feeding experiments with ruminant livestock." *Applied Animal Behaviour Science* 140 (3–4): 105–120. https://doi.org/10.1016/j.applanim.2012.04.008.

Meusnier, I., G. A. C. Singer, J. F. Landry, D. A. Hickey, P. D. N. Herbert, and M. Hajibabaei. 2008. "A universal DNA mini-barcode for biodiversity analysis." *BMC Genomics* 9: 214. https://doi.org/10.1186/1471-2164-9-214.

Monadjem, A. 1997. "Stomach contents of 19 species of small mammals from Swaziland." *South African Journal of Zoology* 32 (1): 23–26.

Montgomery, S. S. J., and W. I. Montgomery. 1990. "Intrapopulation variation in the diet of the wood mouse *Apodemus sylvaticus*." *Journal of Zoology* 222: 641–651.

Morrant, D. S., and S. Petit. 2012. "Strategies of a small nectarivorous marsupial, the western pygmy-possum, in response to seasonal variation in food availability." *Journal of Mammalogy* 93 (6): 1525–35. https://doi.org/10.1644/12-MAMM-A-031.1.

Murray, B. R., C. R. Dickman, C. H. S. Watts, and S. R. Morton. 1999. "The dietary ecology of Australian desert rodents." *Wildlife Research* 26: 421–427.

Newsome, S. D., C. Martinez del Rio, S. Bearhop, and D. L. Phillips. 2007. "A niche for isotope ecology." *Frontiers in Ecology and the Environment* 5 (8): 429–436. https://doi.org/10.1890/060150.01.

Norris, K. H., R. F. Barnes, J. E. Moore, and J. S. Shenk. 1976. "Predicting forage quality by infrared replectance spectroscopy." *Journal of Animal Science* 43 (4): 889–897. https://doi.org/10.2527/jas1976.434889x.

Oguge, N. O. 1995. "Diet, seasonal abundance and microhabitats of *Praomys* (*Mastomys*) *natalensis* (Rodentia: Muridae) and other small rodents in a Kenyan sub-humid grassland community." *African Journal of Ecology* 33: 211–223. https://doi.org/10.1111/j.1365-2028.1995.tb00798.x.

Ozaki, S., C. Fritsch, B. Valot, F. Mora, T. Cornier, R. Scheifler, et al. 2018. "Does pollution influence small mammal diet in the field? A metabarcoding approach in a generalist consumer." *Molecular Ecology* 27 (18): 3700–3713. https://doi.org/10.1111/mec.14823.

Parnell, A. C., R. Inger, S. Bearhop, and A. L. Jackson. 2010. "Source partitioning using stable isotopes: Coping with too much variation." *PLoS ONE* 5 (3): e9672. https://doi.org/10.1371/journal.pone.0009672.

Patel, S., J. Waugh, C. D. Miller, and D. M. Lambert. 2010. "Conserved primers for DNA barcoding historical and modern samples from New Zealand and Antarctic birds." *Molecular Ecology Resources* 10 (3): 431–438. https://doi.org/10.1111/j.1755-0998.2009.02793.x.

Perisse, M., C. R. S. Fonseca, and R. Cerqueira. 1989. "Diet determination for small laboratory-housed wild mammals." *Canadian Journal of Zoology* 67: 775–778.

Phillips, D. L. 2012. "Converting isotope values to diet composition: The use of mixing models." *Journal of Mammalogy* 93 (2): 342–352. https://doi.org/10.1644/11-MAMM-S-158.1.

Phillips, D. L., R. Inger, S. Bearhop, A. L. Jackson, J. W. Moore, A. C. Parnell, et al. 2014. "Best practices for use of stable isotope mixing models in food-web studies." *Canadian Journal of Zoology* 92: 823–835. https://doi.org/10.1139/cjz-2014-0127.

Poinar, H. N., M. Hofreiter, W. G. Spaulding, P. S. Martin, B. A. Stankiewicz, H. Bland, et al. 1998. "Molecular coproscopy: Dung and diet of the extinct ground sloth *Nothrotheriops shastensis*." *Science* 281: 402–406.

Pompanon, F., B. E. Deagle, W. O. Symondson, D. S. Brown, S. N. Jarman, and P. Taberlet. 2012. "Who is eating what: Diet assessment using next generation sequencing." *Molecular Ecology* 21 (8): 1931–1950.

Pulliam, H. R. 2000. "On the relationship between niche and distribution." *Ecology Letters* 3 (4): 349–361. https://doi.org/10.1046/j.1461-0248.2000.00143.x.

Putman, R.J. 1984. "Facts from feces." *Mammal Review* 14 (2): 79–97. https://doi.org/10.1111/j.1365-2907.1984.tb00341.x.

Riaz, T., W. Shehzad, A. Viari, F. Pompanon, P. Taberlet, and E. Coissac. 2011. "ecoPrimers: Inference of new DNA barcode markers from whole genome sequence analysis." *Nucleic Acids Research* 39: e145. https://doi.org/10.1093/nar/gkr732.

Robb, G. N., A. Harrison, S. Woodborne, and N. C. Bennett. 2016. "Diet composition of two common mole-rat populations in arid and mesic environments in South Africa as determined by stable isotope analysis." *Journal of Zoology, London* 300 (4): 257–264. https://doi.org/10.1111/jzo.12378.

Robb, G. N., S. Woodborne, and N. C. Bennett. 2012. "Subterranean sympatry: An investigation into diet using stable isotope analysis." *PLoS ONE* 7 (11): e48572. https://doi.org/10.1371/journal.pone.0048572.

Rodgers, A.R. 1990. "Evaluating preference in laboratory studies of diet selection." *Canadian Journal of Zoology* 68: 188–90. https://doi.org/10.1139/z90-026.

Sang T., D. Crawford, and T. Stuessy. 1997. "Chloroplast DNA phylogeny, reticulate evolution and biogeography of *Paeonia* (Paeoniaceae)." *American Journal of Botany* 84: 1120–1136. https://doi.org/10.2307/2446155.

Sato, J. J., D. Kyogoku, T. Komura, C. Inamori, K. Maeda, Y. Yamaguchi, et al. 2019. "Potential and pitfalls of the DNA metabarcoding analysis for the dietary study of the large Japanese wood mouse *Apodemus speciosus* on Seto Inland Sea islands." *Mammal Study* 44 (4): 221–231.

Sato, J. J., T. Shimada, D. Kyogoku, T. Komura, S. Uemura, T. Saitoh, et al. 2018. "Dietary niche partitioning between sympatric wood mouse species (Muridae: *Apodemus*) revealed by DNA metabarcoding analysis." *Journal of Mammalogy* 99 (4): 952–964.

Schneider, S., R. Steeves, S. Newmaster, and A. S. MacDougall. 2017. "Selective plant foraging and the top-down suppression of native diversity in a restored prairie." *Journal of Applied Ecology* 54 (5): 1496–1504. https://doi.org/10.1111/1365-2664.12886.

Shiels, A. B., C. A. Flores, A. Khamsing, P. D. Krushelnycky, S. M. Mosher, and D. R. Drake. 2013. "Dietary niche differentiation among three species of invasive rodents (*Rattus, R. exulans, Mus musculus*)." *Biological Invasions* 15: 1037–1048. https://doi.org/10.1007/s10530-012-0348-0.

Simon, C., F. Frati, A. Beckenbach, B. Crespi, H. Liu, and P. Flook. 1994. "Evolution, weighing, and phylogenetic utility of mitochondrial gene sequences and a compilation of conserved polymerase chain reaction primers." *Annals of the Entomological Society of America* 87: 651–701.

Soininen, E. M., A. Valentini, E. Coissac, C. Miquel, L. Gielly, C. Brochmann, et al. 2009. "Analysing diet of small herbivores: The efficiency of DNA barcoding coupled with high-throughput pyrosequencing for deciphering the composition of complex plant mixtures." *Frontiers in Zoology* 6: 16. https://doi.org/10.1186/1742-9994-6-16.

Soininen, E. M., L. Zinger, L. Gielly, E. Bellemain, K. A. Bråthen, C Brochmann, et al. 2013. "Shedding new light on the diet of Norwegian lemmings: DNA metabarcoding of stomach content." *Polar Biology* 36: 1069–1076. https://doi.org/10.1007/s00300-013-1328-2.

Soininen, E. M., L. Zinger, L. Gielly, N. G. Yoccoz, J.-A. Henden, and R. A. Ims. 2017. "Not only mosses: Lemming winter diets as described by DNA metabarcoding." *Polar Biology* 40: 2097–2103. https://doi.org/10.1007/s00300-017-2114-3.

Stewart, D. R. M. 1967. "Analysis of plant epidermis in faeces: A technique for studying the food preferences of grazing herbivores." *Journal of Applied Ecology* 4 (1): 83–111.

Stock, B. C., A. L. Jackson, E. J. Ward, A. C. Parnell, D. L. Phillips, and B. X. Semmens. 2018. "Analyzing mixing systems using a new generation of Bayesian tracer mixing models." *PeerJ* 6: e5096. https://doi.org/10.7717/peerj.5096.

Symes, C. T., J. W. Wilson, S. M. Woodborne, Z. S. Shaikh, and M. Scantlebury. 2013. "Resource partitioning of sympatric small mammals in an African forest-grassland vegetation mosaic." *Austral Ecology* 38 (6): 721–729. https://doi.org/10.1111/aec.12020.

Taberlet, P., E. Coissac, F. Pompanon, L. Gielly, C. Miquel, A. Valentini, et al. 2007. "Power and limitations of the chloroplast trnL (UAA) intron for plant DNA barcoding." *Nucleic Acids Research* 35 (3): e14. https://doi.org/10.1093/nar/gkl938.

Taberlet, P., L. Gielly, G. Pautou, and J. Bouvet. 1991. "Universal primers for amplification of three non-coding regions of chloroplast DNA." *Plant Molecular Biology* 17: 1105–1109. https://doi.org/10.1007/BF00037152.

Tate, J. A., and B. B. Simpson. 2003. "Paraphyly of *Tarasa* (Malvaceae) and diverse origins of the polyploid species." *Systematic Botany* 28: 723–737. https://doi.org /10.1043/02-64.1.

Thomas, A. C., B. E. Deagle, J. P. Eveson, C. H. Harsch, and A. W. Trites. 2016. "Quantitative DNA metabarcoding: Improved estimates of species proportional biomass using correction factors derived from control material." *Molecular Ecology Resources* 16 (3): 714–726.

Valentini, A., F. Pompanon, and P. Taberlet. 2009. "DNA barcoding for ecologists." *Trends in Ecology and Evolution* 24 (2): 110–117.

Vaughan, T. A., J. M. Ryan, and N. J. Czaplewski. 2011. *Mammalogy*. 6th ed. Sudbury, MA: Jones and Bartlett Publishers.

Vieira, M. V. 2003. "Seasonal niche dynamics in coexisting rodents of the Brazilian Cerrado." *Studies on Neotropical Fauna and Environment* 38 (1): 7–15. https://doi .org/10.1076/snfe.38.1.7.14034.

Vieira, M. V., R. Finotti, and R. T. Santori. 2015. "From simple questions to complex answers: A research program based on diet selection and water balance of small mammals." *Oecologia Australis* 19 (1): 32–46. https://doi.org/10.4257/oeco.2015.1901.03.

White, T. J., T. D. Bruns, S. B. Lee, and J. W. Taylor. 1990. "Amplification and direct sequencing of fungal ribosomal RNA genes for phylogenetics." In *PCR Protocols: A Guide to Methods and Applications*, edited by M. A. Innis, D. H. Gelfand, and J. J. Sninsky. San Diego: Academic Press.

Wirminghaus, J. O., and M. R. Perrin. 1992. "Diets of small mammals in a Southern African temperate forest." *Israel Journal of Zoology* 38 (3–4): 353–361.

Zeale, M. R., R. K. Butlin, G. L. Barker, D. C. Lees, and G. Jones. 2011. "Taxon-specific PCR for DNA barcoding arthropod prey in bat faeces." *Molecular Ecology Resources* 11: 236–244.

12. Small Mammal Behaviors: Personality, Activity, and Fear

Background

Behavior is most often defined as a response or series of responses to a stimulus. This simple definition belies the complexity associated with the many ways animals may respond to stimuli. For example, two stimuli may interact to affect an animal's response. Consider an animal responding to the stimulus of hunger. The logical response to this stimulus is to forage. However, a second stimulus, presence of a predator, may result in the response of hiding. Here the second stimulus was more important and had the greater influence on behavior. However, if hunger was greater and predation risk was reduced, we may expect to see vigilant foraging such that the animal forages but scans for predators. As the number of simultaneous stimuli increases, a myriad of competing responses present themselves. The difficulty of studying behavior is further compounded in that all animals of a given species will not exhibit the same response to the exact same set of stimuli.

Different responses to an identical set of stimuli by individuals of the same species, age, sex, and living in the same general environment arise due to differences in personality (Roche et al. 2016). Thus, in the context of studying small mammals, behavior can be considered as anything a given organism does; individual differences in behavior, then, can be considered the result of organism personality.

Animal behaviors provide the underpinnings for studying many aspects of the ecology of free-ranging animals. In many cases, the behavior under study is a response to a particular environmental stimulus. For example, animal movements may be considered as behavioral response to habitat fragmentation (Diffendorfer et al. 1995) or food abundance (Morris et al. 2011). Habitat selection itself is often considered a behavioral response to resource needs relative to resource availability and risk (Brown and Kotler 2004). Likewise, activity patterns may be studied as a behavioral response to weather (Vickery and Bider 1981), temperature at a particular location within a burrow (Kenagy 1973), resource acquisition (Brown 1956), or risk of predation (Lima and Dill 1990, Jacob and Brown 2000). In other cases, researchers are interested in how behaviors or personality affect other ecological phenomena. For example, an animal's personality may affect its response to predators (Cremona et al. 2015), survival (Rödel et al. 2015), reproductive success (Le Cœur et al. 2015), and probability of parasitic infection (Boyer et al. 2010); animal personality may also influence how foraging behaviors affect plant communities (Cherry et al. 2016).

Aims

In this chapter we introduce methods for studying aspects of animal personality and behavior. Importantly, this chapter is not an exhaustive treatment of the topic, as this is simply not possible in a single chapter. Instead, we first focus on personality measurements, particularly measures of boldness or exploration, because personalities of small mammals may be useful in better quantifying other behaviors (Roche et al. 2016). We also present techniques for measuring when or under what conditions an animal is active, assuming that researchers interested in measuring activity patterns will provide the context for assessing activity. We do not discuss development of holistic activity budgets, as their development may only be possible by visually observing or otherwise recording animal behaviors during the periods of interest. We also introduce a generalized approach for measuring perceived risk and developing a landscape of fear—a map depicting areas on a gradient from safe to risky as perceived by the animal. We conclude with a brief discussion of novel tools for measuring behaviors and potential application of using animal behaviors as leading indicators of population well-being.

Approaches

Animal Personality

Animal ecologists are likely aware of the often striking differences in behavior that animals of a given species may exhibit in response to a specific stimulus. These differences arise because individuals have different personalities. Although the existence of animal personalities is well recognized, incorporating measures of personality into ecological field studies is a relatively recent endeavor (Sih et al. 2004, Boyer et al. 2010, Montiglio et al. 2012, Le Cœur et al. 2015).

Importantly, for a given behavior to be considered a personality trait, the behavior must also be repeatable (i.e., the individual responds to the same stimulus in a similar way in repeated trials). In studies focused on identifying personality traits, repeated behavioral measurements are taken on multiple individuals, and individual-specific differences and repeatability of behaviors are both required to demonstrate personality (Chock et al. 2017). In studies attempting to use personality to better understand ecological phenomena, the rigor needed to demonstrate personality traits may not be feasible.

Surprisingly few basic approaches to assaying animal personality have been developed, but variations of these basic techniques are numerous. Unfortunately, many of these are not well suited for use in a field setting. As a result, we focus our discussion of personality assays on those techniques that are easily implemented on small mammals in the field. Further, in the most general sense, any behavior that is both consistent within an individual but differing among individuals can be considered as a personality trait (Réale et al. 2010). Rather than attempt to cover such a particularly broad spectrum of potential personalities, we focus on personality traits that seem relatively common in studies of small mammals; these are assays of boldness and/or exploration.

Boldness is commonly defined as an individual's response to a risky environment, whereas exploration deals with an individual's reaction to a novel situation (Réale et al. 2007, Chock et al. 2017). Although it is possible that a bold animal may not be exploratory and vice-versa, we discuss these two personality traits as a behavioral syndrome, i.e., a correlated suite of behavioral traits (Sih et al. 2004). Thus, we discuss boldness and exploratory personality traits as though they are strongly and positively related. When providing applications of how boldness and exploration have been measured and used in ecological studies, we also point out where these empirical studies revealed evidence for existence of the behavioral syndrome.

Perhaps one of the simpler measures of boldness in small mammals is an animal's willingness to enter a trap (Le Cœur et al. 2015). The simplicity of this approach, when coupled with studies that require animal capture, may lead a researcher to consider an animal's willingness to enter a trap as a particularly attractive way to measure boldness. However, this approach can be fraught with problems. First, there must be individual heterogeneity with regard to trappability, which means that trapping intervals need to be sufficiently long to offer opportunity for recaptures. Further, numerous trapping intervals are needed to show that willingness to enter a trap is repeatable among individu-

Fig. 12.1. Handle bag test conducted on a cotton rat (*Sigmodon hispidus*) at The Jones Center at Ichauway, Georgia

als. For example, an animal that was deemed highly trappable relative to another during one particular trapping interval should remain more trappable in subsequent intervals. Relatively long life spans increase suitability of this particular assay, unfortunately this is generally not the norm for many small mammal species.

Another easily performed field assay of personality that is particularly applicable to small mammals is the handling bag test (Fig. 12.1). To implement this test, researchers place recently captured animals in a small mesh handling bag and then suspend the bag in the air for one minute while recording the time the animal remains immobile (Martin and Réale 2008). Increased immobility time is interpreted as increased docility. Interestingly, docility was uncorrelated with exploration behavior in this study, suggesting no evidence for a behavioral syndrome. In another study (Montiglio et al. 2012), a strong negative correlation between exploration and docility was detected. Both studies used the eastern chipmunk (*Tamias striatus*) as the study animal.

Boldness has also been measured using a variety of approaches in which the animal is placed in a novel environment and its behavior measured as an assay of boldness (Belzung 1999). The least elaborate of these techniques is

Fig. 12.2. Hole-board test area being used to measure boldness of a fox squirrel (*Sciurus niger*) (Photo: Catherine Frock)

the open field test. This procedure involves placing the small mammal into a novel arena and then measuring its behavior for a set period of time (Walsh and Cummins 1976). Rodents, when placed in an open field arena, will tend to spend time on the periphery of the arena. Thus, taking less time to leave the periphery or spending more time in central regions of the open field can be indicative of a bolder personality.

Perhaps the most-used method for measuring boldness is one that uses a hole-board (Fig. 12.2). The hole-board apparatus consists of a wooden board with holes drilled in grid fashion. This board serves as the cover for a box such that there is a space below the hole-board. The space between the top (with holes) and the floor of the box is small, such that the animal being studied cannot fit between the top and bottom of the box, but it can dip its head into the holes. The animal of interest is placed on the hole-board for a fixed period of time and the number of head dips (i.e., placement of the head in a hole) is recorded. Greater frequencies of head dipping are considered more indicative of boldness. Similar to the open field test, hole-board assays generally take place in a laboratory setting but use in a field setting should be easily implemented.

A number of additional tests have been suggested for measuring boldness, and some of these tests (e.g., the y-maze and elevated plus maze) are popular within laboratory settings (see examples in Belzung 1999). Although adapting these procedures to a field setting, as suggested using the hole-board and open field assays, may be possible, the physical size of mazes is large, and using them to collect behavioral data is time-consuming. As a result, these approaches may be best restricted to a laboratory setting.

Applications

Boyer et al. (2010) used a hole-board test and trappability to assay boldness in chipmunks (*Tamias sibiricus*). They found that trappability and exploration of the hole-board were positively correlated and that that these measurements were correlated with space use—bolder animals had larger home ranges. Increased space use was also correlated with tick burden, demonstrating that chipmunk personality affected space use and parasite burden in this species.

Montiglio et al. (2012) studied eastern chipmunk personality relative to long-term stress reactivity. In this study, eastern chipmunk personality assays were conducted using trappability, handling bag tests, and open field tests. Personality traits were then compared to fecal hormone levels and heart rate. The authors identified three aspects of chipmunk personality: exploration, vigilance, and centrality (i.e., the proportion of time chipmunks spent in the center of the open field arena—often considered a measure of boldness). Docility was negatively correlated with exploration. There was no correlation between fecal hormone levels and exploration, but there was a relationship between variability in hormone levels and exploration. Animals with less variable cortisol levels tended to have more exploratory personalities.

Chock et al. (2017) observed behaviors of degus (*Octodon degus*) captive within wire-mesh traps and used these data to test the hypothesis that exploratory and boldness traits represented a behavioral syndrome in this species. They assessed exploratory behavior by considering the trap a novel environment and scoring animals based on whether they moved around within the trap during a fixed observation period. They assessed boldness by touching the animal on the hind quarters with a pencil eraser; turning to charge the pencil was considered indicative of boldness. The researchers found that exploratory behavior and boldness were correlated in degus, suggesting presence of a behavioral syndrome. Despite evidence for a behavioral syndrome, the authors noted significant negative association between exploratory traits and burrow sharing. Exploratory animals tended to share burrows with non-exploratory animals. There was no relationship between boldness measures and burrow sharing.

Melange et al. (2016) measured personality traits of the invasive hairy-tailed bolo mouse (*Necromys lasiurus*) to determine whether there was evidence for a behavioral syndrome consisting of activity, exploration, and neophilia personality traits. They used an open field test, a modified open field test, and a hole-board to measure personality traits. The authors used the results of the personality assays to discuss potential implications of personality on the success of invasion.

Korpela et al. (2011) measured exploratory behavior in bank voles (*Myodes glareolus*) by placing the vole in an open arena with the floor gridded into 24 cells. Wheat flour was sprinkled on the floor of the arena. A vole was placed in the arena and allowed to move about for 10 minutes with the lights turned off. The number of cells the vole visited was determined by examining its tracks in the flour. Number of cells visited (1–24) was recorded for each vole as an exploration score. This test was used with other personality assays to identify three bank vole personalities and to determine if bank vole density affected personality traits.

Animal Behaviors

Activity Patterns

Animal activity patterns are often intrinsic; for example, some species are largely nocturnal, others are diurnal, others crepuscular, and still others may have no discernable peaks in activity. At the same time, activity patterns generally exhibit some degree of plasticity such that dominant patterns may be overridden by environmental factors. In studies of small mammals, researchers are often interested in both aspects, with some studies seeking to elucidate the intrinsic norm, while others seek to understand how environmental influences may affect activity patterns (Kenagy 1973, Scheibler et al. 2014).

Here, we only consider two activity states, either active or inactive, and make no distinction among potential behaviors that could fall outside these two broad categories. For example, an active animal may be traveling between a refuge site and a foraging site, it may be actively foraging, fleeing a predator, or seeking a mate; however, approaches we discuss here will only reveal if the animal is active. Techniques that seek to identify actual behaviors when an animal is active or inactive require visual monitoring for accurate assessment and are of limited suitability when studying small mammals in a field setting.

In some cases, more precise behaviors can be inferred without visual observation. For example, an animal deemed motionless within its burrow may be generally considered as resting. Likewise, if an animal is deemed active at a known food source, it is likely that it is foraging. When we discuss examples of studies that measured small mammal activity, we point out when the authors made such assumptions.

One of the most obvious approaches to monitoring small mammal activity is through use of radio transmitters. Radio-tagged animals can be tracked and their movements determined (Schradin 2006); increased movement is indicative of increased activity (Chapter 8). In addition to intensive tracking as a proxy for activity, radio transmitters can be equipped with motion sensors that transmit a different signal based on whether the animal is moving or whether the animal has moved during a predetermined time frame. These transmitters, when coupled with a telemetry receiver and a remote data logger, can collect data to enable quantification of individual activity and permit construction of an activity pattern distribution associated with the monitored population (Kays et al. 2011).

Times of capture can also be used to construct crude activity patterns. If labor permits, traps can be run frequently, permitting reasonable estimates of the time of capture (Kenagy 1973). When the researcher processes the captured animal, the time of capture can be determined. Live traps also can be equipped with a timer that triggers when the trap closes (Sutherland and Predavec 2010). Multiple captures allow researchers to calculate an estimate of when animals are active or inactive. However, only in rare cases will individuals be captured with sufficient frequency as to permit construction of individual-based activity patterns.

Similarly, camera trapping (Fig. 12.3) can be used to identify activity patterns by recording time of image acquisition (Pratas-Santiago et al. 2016, So-

Fig. 12.3. Camera trap used to measure activity at a small mammal foraging tray

vie et al. 2019). Like physical trapping, this approach may not be successful for calculating individual-based activity patterns because there may be too few individual captures or because individuals cannot be readily identified. However, if individual identification is possible, perhaps through uniquely marking animals, and camera trapping intensity is adequate, derivation of activity rates for individuals may be possible.

Other researchers have used passive integrated transponder (PIT) tags (Smyth and Nebel 2013, Scheibler et al. 2014) (see Chapter 5) to detect animals when they pass by a detector. This permits detection of activity associated with individual animals and potentially individual activity patterns. Although use of PIT tags can be adapted to a variety of situations, they are perhaps most useful for measuring when an animal leaves and reenters a refuge site or when they are using a particular foraging patch.

In some cases, animal tracks can be used to assess activity. When using this approach, a tracking substrate is placed in an area where small mammals may frequent. The substrate is examined after a period of time, animal tracks recorded, and the substrate wiped clean in preparation for the next bout of sampling. The number of animals leaving signs in the substrate is then used to infer activity. Using animal tracks to assess activity patterns is likely best left to situations where precision regarding timing of activity is not critical. For example, activity as it relates to moon phase may be successfully investigated using animal tracks, but if hourly estimates of activity are needed, other methods would be preferred. With some species, it may be difficult to distinguish between species as small mammal tracks can be difficult to distinguish in tracking substrate (Vickery and Bider 1981). Use of tracking tubes (Maybee 1998) or artificial foraging patches (Brown 1988) should also prove acceptable as a means of detecting tracks (Chapter 2).

Applications

Mark-recapture techniques were used to quantify nighttime activity patterns of three heteromyid rodents (Kenagy 1973). Traps were run at intervals of 2–2.5 hours during the night, and time-specific frequency of captures were used to measure activity patterns of each species. Underground activity was also measured by tagging animals with a radioactive gold wire, and the depth of animals and location in the burrow was estimated by measuring gamma-ray emissions. Animals were active in their burrows throughout the day, and location within the burrow was influenced by temperature in the burrow, whereas aboveground activity was influenced by ambient temperature (Kenagy 1973).

O'Farrell (1974) used mark-recapture analyses to measure activity patterns of 12 desert rodent species during a one-year period. Traps were checked every two hours, and data were used to determine whether rodents were active year-round (i.e., did not hibernate), were seasonally active, or sporadic in occurrence (i.e., nomadic species). He also evaluated daily activity and correlated animal activity with environmental conditions.

Although data were presented with the primary intent of providing proof of concept, attaching timers to traps provided capture times associated with 136 captures of Australian bush rats (*Rattus fuscipes*). These capture times were useful for determining diel activity patterns for this species, and presence of the timer did not result in biased capture rates (Sutherland and Predavec 2010).

Desert hamsters were marked with passive transponders, and microchip readers were placed at their burrows. Movement in and out of burrows was recorded, and these data were used to measure time in and out of the burrow relative to moon phase, ambient temperature, and soil temperature, all of which influenced activity (Scheibler et al. 2014).

Activity of ocelots (*Leopardus pardalis*) and their small mammal prey was estimated using camera traps. Researchers obtained over 1500 images from camera traps and used these images and their associated time stamps to estimate activity patterns of both predator and prey, and to assess temporal overlap in activity. They noted that, during full moon periods, activity of ocelots overlapped activity of nocturnal prey significantly more than other prey (Pratas-Santiago et al. 2016).

Vickery and Bider (1981) established a 0.45 m wide by 91.5 m long transect of sand perpendicular to a forest edge. The transect was covered by a plastic canopy to prevent rain from obscuring tracks. They checked the transect each morning to observe and count small mammal tracks. The authors visited the transect 651 times over a seven-year period and found that rodent activity was greatest on warm rainy nights.

Schradin (2006) used radiotelemetry to determine activity of striped mice (*Rhabdomys pumilio*) in open habitats in South Africa. Once an animal was located, direct observations were used to record specific diurnal activities and social interactions. Striped mice moved most during mornings and evenings and rested in bushes during the heat of the day.

Predation Risk

Most species of small mammal serve as prey for other animals. Thus, risk of predation constantly weighs on their behavioral decisions. This is as it should be, as an underestimate of this risk has profound and irreversible consequences. At the same time, however, overestimation of this risk has consequences, such as reduced foraging opportunities and reduced fitness (Brown 1988). As a result, there is strong selective pressure on prey species to respond to risk of predation appropriately. Understanding the importance of the food/risk trade-off provides much opportunity for studying small mammal ecology.

Brown (1988) presented the theoretical underpinnings and empirical evidence suggesting that the density of food left when a forager abandons a feeding patch—the giving up density (GUD)—could be used to assess perceived predation risk. Optimal foraging theory was essential in the development of the GUD concept. An animal that forages optimally should harvest food from a patch such that the value of food consumed pays for the metabolic opportunity cost (C) of foraging, the predation cost of foraging (P), and the missed opportunity costs (MOC) of foraging. If food is considered a depletable resource within a patch, the harvest rate of food (H) will decrease the longer the animal forages. When H no longer pays for the cost of C, P, and MOC, the animal should leave the foraging patch. The food remaining in the patch is the GUD. When MOC and C are controlled, only P remains as the predation cost of foraging. Controlling C and MOC is accomplished by providing similar patches to the same set of animals under the same microclimatic conditions and the same time (Brown 1988), i.e., providing temporal and spatial replicates that differ only in terms of predation risk within the area being studied.

Importantly, use of GUDs is not restricted to predation-risk studies alone. In their review of GUDs, Bedoya-Perez et al. (2013) discussed modifications to the basic model proposed by Brown (1988). These modifications allowed researchers to study effects of interference competition (Kotler and Brown 1988), presence of plant toxins (Schmidt 2000), benefits of water in a desert setting (Schrader et al. 2008), risk of injury during foraging (Berger-Tal et al. 2009), and benefits of having information while foraging (Olsson and Brown 2010). Thus, GUDs have diverse application for studying ecological factors that affect foraging outcomes, and a basic understanding of their application can be an important tool when studying small mammals.

Many methods and sampling techniques are relatively standardized for small mammals. Unfortunately, use of foraging patches for quantifying GUDs is much less standardized. For example, there are no standard traps, tags, or transmitters, such as may be used in other studies. Thus, researchers must use the guiding principles discussed below, and trial and error, to develop foraging patches that work. Once this is done, use of the patch to calculate GUDs and address questions of ecological importance is only limited by researcher creativity and funding. Although GUDs have been successfully calculated without using artificial foraging patches (Morris 2005), use of artificial patches is the norm, with most patches consisting of a mixture of food and an inedible substrate in a container that is accessible to the target species.

When developing an operational foraging patch, researchers must establish a combination of food item, food amount, inedible foraging substrate, and substrate volume such that the target species readily consumes food from the patch but does not completely deplete food in the patch. Importantly, simply providing an excess of food will not provide a usable GUD, as the harvest rate of food will not decline the longer an animal forages. Instead, the animal must work for the food so that each bit of food consumed forces the animal to work harder for the next bit.

Choice of food used in GUD experiments should be something that is readily consumed by the animal being studied. Finding candidate food items should be possible using a literature review. Brown (1988) used millet seeds in his original work of granivorous rodents, which he used to introduce the technique. Millet seeds have also been used to measure GUDs of herbivorous rodents (Darracq et al. 2016). Use of grains has the added bonus of being relatively easy to quantify at the end of the foraging bout (see below), but other food items can be used as needed. Additionally, more desirable foods result in being able to use less food in the foraging patch.

Once a potential food item is settled upon, an inedible substrate must be chosen in which the food is mixed. Sifted sand (Van Der Merwe and Brown 2008, Orrock and Fletcher 2014, Darracq et al. 2016) or soil (Brown 1988) is commonly used as a foraging substrate for rodents. The purpose of the foraging substrate is to ensure harvest of food from the patch becomes more difficult as the food items are consumed. In the above-mentioned studies, 0.5–4 L of sand or soil were used in all cases. Amount of food provided ranged from 2–8 g in all cases. Once a food and substrate are chosen, the researcher must determine the appropriate ratio to ensure food is discovered and consumed but not to completion. Determining the ratio of food to foraging medium is often a matter of experience and trial and error.

As a practical consideration, those wishing to use GUDs in their research should consider how food and substrate are to be separated to determine quantity of remaining food. Some potential food items (e.g., alfalfa pellets) can disintegrate over time or when they get wet, rendering a particular foraging bout useless. For this reason, we suggest use of grains as food when possible, as grain holds up well to the elements and can be readily separated from most substrates by sifting. Additionally, users should consider placing some sort of cover over the foraging patch to protect the patch from rain.

Although there may be many approaches to getting animals to forage in a foraging patch, and for arriving at a working ratio of substrate to food, we offer the following advice for developing a foraging patch used to calculate small mammal GUDs. After deciding on a food and substrate, select a container that will ultimately contain the food-substrate mixture that will be easily ac-

Fig. 12.4. Peromyscus sp. (A) and cotton rat (Sigmodon hispidus) (B) utilizing foraging trays designed to estimate giving up densities (GUDs)

cessible by the target species (Fig. 12.4). Place containers and substrate in the area to be sampled but do not mix the food into the substrate. Instead, place the food on top of the substrate such that the foraging animals can readily find and consume the food. Check foraging patches daily and note whether food was consumed by the species of interest; a well-placed trail camera can facilitate identification if tracks are not discernible in the substrate. Once foraging patches are regularly being visited and food readily consumed, measure a quantity of food and mix it thoroughly in the substrate. Continue to check patches daily, but now separate substrate from food and measure quantify of food remaining. If there was clear evidence that the target species consumed the food yet some food was left behind, a working foraging patch has been developed. Note that not all foraging patches need to be visited in the initial study area, as some of the patches may be placed in microhabitats with great predation risk; thus, they may be avoided by the target species. When placing patches in a new area, it is a good idea to leave food on top of the substrate for a few days to ensure small mammals associate the foraging patch with food. After this association has occurred within a given area, food can be mixed with the substrate and GUDs can be calculated. After foraging patches are producing usable GUDs, duration of monitoring in a given area is often brief, ranging from one day to a week (Brown 1988, Druce et al. 2006, Orrock and Fletcher 2014), but duration of monitoring should ultimately match study objectives.

Perhaps one of the more interesting approaches to using GUDs is to develop a landscape of fear (Van Der Merwe and Brown 2008). Laundre et al. (2001) argued that animals sacrifice foraging opportunities to reduce their risk of predation, such that animals match their behavior to perceived predation risk as they move across the physical landscape. Thus, Laundre et al. (2001) postulated that it may be useful to envision a second landscape—a landscape of fear—in which prey perceive peaks and valleys based on differing levels of predation risk.

Using GUDs to develop a landscape of fear involves placing working foraging patches in a grid fashion across the landscape of interest. Foraging patches are checked, and GUDs for each station are calculated. A map of the landscape of fear can then be constructed using spatial statistics such as a spline curve to spatially interpolate contours such that GUDs are represented as contour lines (Van Der Merwe and Brown 2008)

When developing a map of the landscape of fear, space between foraging patches should be carefully considered because predation risk can often vary at extremely small scales. For example, if a single shrub represents a safe environment (Loggins et al. 2019), placing foraging patches at 20 m spacing will likely miss much of the variation in predation risk within the sampled grid. Obviously, this spacing will be related to body size and home range of the small mammal and vegetation heterogeneity in the area being sampled (Loggins et al. 2019). Although not employed by Van Der Merwe and Brown (2008), pilot studies to investigate different spacing between foraging patches may elucidate the scale that maximizes GUD variation.

We warn that our simple explanation of developing a foraging patch and then using it to calculate GUDs is not as simple as it seems. Researchers attempting to use GUDs in field research invariably experience unforeseen difficulties, and perseverance will be key to successful applications. Indeed, Bedoya-Perez et al. (2013) echo this idea and provide a number of problems that researchers should consider when working with GUDs. Their work should be consulted when designing studies that rely on GUDs.

Additionally, recent work by McMahon et al. (2018) suggests that nutritional preferences and predation risk can covary. Although their study was carried out using grasshoppers, the results warrant mention. When grasshoppers were provided a protein-rich diet, GUDs increased with increasing predation risk, as predicted by current theory. However, when grasshoppers experience predation risk, they prefer carbohydrate-rich food. As a result, grasshoppers on a carbohydrate-rich diet did not exhibit increased GUDs as predation risk increased. Thus, if predation risk increases prey nutritional preference for a given food item, and that item is used in foraging patches, predation risk and nutrition could be confounded in analyses. It is unknown the degree this work will generalize to studies of small mammals, but the results should be considered when selecting a food for use in foraging patches intended to provide GUDs.

Applications

When Brown (1988) introduced the concept of GUDs as a measure of predation risk, his foraging patches consisted of aluminum tray containers filled with 3 L of sifted soil as a substrate. In this substrate he mixed 3 g of millet seeds. His work focused on calculating GUDs for four granivorous rodents. He found that GUDs consistently differed based on forager species, microhabitat, date, and station.

In a designed experiment to evaluate the effects of red imported fire ants (*Solenopsis invicta*) on hispid cotton rat (*Sigmodon hispidus*) foraging behavior, Darracq et al. (2016) used a foraging patch consisting of a plastic pan with 2 L of sifted sand as a substrate and 2 g of millet seed for food. Cotton rats foraging in sites with red imported fire ants had greater GUDs than sites where fire ants had been experimentally reduced.

Orrock and Fletcher (2014) studied deer mice (*Peromyscus maniculatus*) foraging relative to island fox (*Urocyon littoralis*) abundance. Foraging patches consisted of round plastic containers with translucent sides filled with 0.47 L of sifted sand substrate containing approximately 4.7 g of millet for food. They found that deer mouse GUDs were positively associated with island fox abundance on the island but that GUDs were also affected by risky and safe microhabitats and risky and safe foraging nights.

Kotler et al. (2010) studied the response of Allenby's gerbils (*Gerbilus andersoni allenbyi*) to moonlight while under the threat of red fox (*Vulpes vulpes*) predation using GUDs. Their foraging patches used a plastic tray for a container, 3 L of sifted sand for substrate, and 3 g of millet seeds as food. They observed greater GUDs on moonlit nights.

Druce et al. (2006) studied foraging behavior of rock hyraxes (*Procavia capensis*) to study how environmental factors affected foraging. Foraging patches consisted of a circular plastic tray using a 2 L substrate of sand and 40 g of alfalfa pellets. They found that both spatial and temporal foraging patterns were likely more influenced by predation risk than other habitat factors.

Van Der Merwe and Brown (2008) developed maps of the landscape of fear for Cape ground squirrels (*Xerus inauris*). Foraging patches consisting of 3 L plastic trays filled with a 2 L substrate of sand and 8 g of wheat was placed in an 8 × 8 grid with 10 m between patches. They divided landscapes of fear into three categories (high, medium, and low) representing the predation cost of foraging and estimated between 31% and 92% of sampled areas represented high foraging costs.

Conclusions and Future Direction

Scientific understanding of the importance of animal personality and behavior to survival and fitness, and ultimately of the impacts of animal personality and behavior on whole ecosystems, is in its infancy. If for no other reason than to contribute data toward increased understanding of the effects of animal behavior on broader ecological processes, small mammal ecologists should consider incorporation of behavioral measurements as core components of research projects.

Conservation monitoring is increasingly important for managing and restoring imperiled wildlife. Unfortunately, most monitoring relies on trailing indicators of population dynamics. In other words, monitoring tells how the past has affected population dynamics. When managing species of conservation concern or game species that are harvested by humans, leading indicators of population processes may allow management intervention before undesirable population consequences are incurred. Kotler et al. (2016) suggest three classes of behavioral indicators that may have particular promise as leading indicators of population well-being: diet, patch use, and habitat selection. With regard to diet, they note that healthy populations often exhibit seasonal dietary fluctuations, but departures from the normal seasonal diet may be indicative of a population experiencing a lack of food. Likewise, patch use and habitat selection offer promise as leading indicators because animals forage first in habitats that are in or near safe areas. As a result, foraging patches near refuge sites are depleted before more distant sites are depleted. Depletion of distant foraging patches may be indicative of decreased predation risk or of declining population health. Kotler et al. (2016) go on to suggest that measuring GUDs over time and among patches may provide valuable insights into population well-being.

One current limitation of behavioral research on small mammals is the ability to provide a stimulus and subsequent monitoring of behavioral responses of the animal in a field setting. One recent advance is the automated behavioral response (ABR) system (Suraci et al. 2017), which combines trail cameras with an acoustic broadcast device. When an animal triggers the system, the animal's behavior is video-recorded for a period of time before and after a sound is broadcast. This enables researchers to examine behavioral response of prey to different predators. The applications of ABRs to small mammal studies will enable investigation into how different predators affect prey behavior within different habitats and permit insight into how predator community composition may affect prey habitat selection and habitat-specific vigilance.

Bluetooth technology has promise for estimating animal activity patterns, assessing when monitored animals are in proximity to one another, and for evaluating animal presence at specific locations, such as a GUD foraging station. Very small Bluetooth loggers can be attached to small mammals and used to monitor interactions with other small mammals, and to monitor presence in a specific area. When monitored animals are in close proximity (approx. 1 m) to each other or to a stationary logger, the interaction and a time-date stamp are recorded and stored to the logger. This information can then be downloaded when the animal comes in contact with a remote base station or the animal is recaptured (Berkvens et al. 2019).

REFERENCES

Bedoya-Perez, M. A., A. J. R. Carthey, V. S. A. Melia, C. McArthur, and P. B. Banks. 2013. "A practical guide to avoid giving up on giving-up densities." *Behavioral Ecology and Sociobiology* 67: 1541–1553.

Belzung, C. 1999. "Measuring rodent exploratory behavior." In *Handbook of Molecular-Genetic Techniques for Brain and Behavioral Research,* edited by W. E. Crusio and R. T. Gerlat, 738–449. Amsterdam: Elsevier Science.

Berger-Tal, O., S. Mukherjee, B. P. Kotler, and J. S. Brown. 2009. "Look before you leap: Is risk of injury a foraging cost?" *Behavioral Ecology and Sociobiology* 63: 1821–1827.

Berkvens R., I. H. Olivares, S. Mercelis, L. Kirkpatrick, and M. Weyn. 2019. "Contact detection for social networking of small animals." In *Advances on P2P, Parallel, Grid, Cloud and Internet Computing. 3PGCIC 2018. Lecture Notes on Data Engineering and Communications Technologies*, vol. 24, edited by F. Xhafa, F. Y. Leu, M. Ficco, and C. T. Yang. Cham, Switzerland: Springer. https://doi.org/10.1007/978-3-030-02607-3_37.

Boyer, N., D. Réale, J. Marmet, B. Pisanu and J. L. Chapuis. 2010. "Personality, space use, and tick load in an introduced population of Siberian chipmunks *Tamias sibiricus*." *Journal of Animal Ecology* 79: 538–547.

Brown, J. S. 1988. "Patch use as an indicator of habitat preference, predation risk, and competition." *Behavioral Ecology and Sociobiology* 22: 37–47.

Brown, J. S., and B. P. Kotler. 2004. "Hazardous duty pay and the foraging cost of predation." *Ecology Letters* 7: 999–1014.

Brown, L. E. 1956. "Field experiments on the activity of the small mammals *Apodemus*, *Clethrionomys*, and *Microtus*." *Proceedings of the Zoological Society of London* 126: 549–564.

Cherry, M. J., R. J. Warren, and L. M. Conner. 2016. "Fear, fire, and behaviorally mediated trophic cascades in a frequently burned savanna." *Forest Ecology and Management* 368: 133–139.

Chock, R. Y., T. W. Wey, L. A. Ebensperger, and L. D. Hayes. 2017. "Evidence for a behavioral syndrome and negative social assortment by exploratory personality in the communally nesting rodent, *Octodon degus*." *Behavior* 154: 541–562.

Cremona, T. A., V. S. A. Mella, J. K. Webb, and M. S. Crowther. 2015. "Do individual differences in behavior influence wild rodents more than predation risk?" *Journal of Mammalogy* 96: 1337–1343.

Darracq, A. K., L. M. Conner, J. S. Brown, and R. A. McCleery. 2016. "Cotton rats alter foraging in response to an invasive ant." *PLoS ONE* 11 (9): e0163220.

Diffendorfer, J. E., M. S. Gaines, and R. D. Holt. 1995. "Habitat fragmentation and movements of three small mammals (*Sigmodon*, *Microtus*, and *Peromyscus*)." *Ecology* 76: 827–839.

Druce, D. J., J. S. Brown, J. G. Castley, G. I. H. Kerley, B. P. Kotler, R. Slotow, and M. H. Knight. 2006. "Scale-dependent foraging costs: Habitat use by rock hyraxes (*Procavia capensis*) determined using giving-up densities." *Oikos* 115: 513–525.

Jacob, J., and J. S. Brown. 2000. "Microhabitat use, giving-up densities and temporal activity as short- and long-term anti-predator behaviors in common voles." *Oikos* 91: 131–38.

Kays, R., S. Tilak, M. Crofoot, T. Fountain, D. Obando, A. Ortega, et al. 2011. "Tracking animal location and activity with an automated radio telemetry system in a tropical rainforest." *The Computer Journal* 54: 1931–1948.

Kenagy, G. J. 1973. "Daily and seasonal patterns of activity and energetics in a heteromyid rodent community." *Ecology* 54: 1201–1219.

Korpela, K., J. Sundell, and H. Ylönen. 2011. "Does personality in small rodents vary depending on population density?" *Oecologia* 165: 67–72.

Kotler, B. P., and J. S. Brown. 1988. "Environmental heterogeneity and the coexistence of desert rodents." *Annual Review of Ecology and Systematics* 19: 281–307.

Kotler, B. P., J. Brown, D. Mukherjee, O. Berger-Tal, and A. Bouskila. 2010. "Moonlight avoidance in gerbils reveals a sophisticated interplay among time allocation, vigilance and state-dependent foraging." *Proceedings of the Royal Society B* 277: 1469–1474.

Kotler, B. P., D. W. Morris, and J. S. Brown. 2016. "Direct behavioral indicators as a conservation and management tool." In *Conservation Behaviors: Applying Behavioral Ecology to Wildlife Conservation and Management*, edited by O. Berger-Tal and D. Saltz, 307–351. Cambridge: Cambridge University Press.

Laundre, J. W., L. Hernández, and K. B. Altendorf. 2001. "Wolves, elk, and bison: Reestablishing the 'landscape of fear' in Yellowstone National Park, U.S.A." *Canadian Journal of Zoology* 79: 1401–1409.

Le Cœur, C., M. Thibault, B. Pisanu, S. Thibault, J. L. Chapuis, and E. Baudry. 2015. "Temporally fluctuating selection on a personality trait in a wild rodent population." *Behavioral Ecology* 26: 1285–1291.

Lima, S. J., and L. M. Dill. 1990. "Behavioral decisions made under the risk of predation: A review and prospectus." *Canadian Journal of Zoology* 68: 619–640.

Loggins, A. A., A. M. Shrader, A. Monadjem, and R. A. McCleery. 2019. "Shrub cover homogenizes small mammals' activity and perceived predation risk." *Scientific Reports* 9: 1–11.

Mabee, T. J. 1998. "A weather-resistant tracking tube for small mammals." *Wildlife Society Bulletin* 26: 571–574.

Martin, J. G. A., and D. Réale. 2008. "Temperament, risk assessment and habituation to novelty in eastern chipmunks, *Tamias striatus*." *Animal Behavior* 75: 309–318.

McMahon, J. D., Lashley, M. A., Brooks, C. P. and Barton, B. T. 2018. "Covariance between predation risk and nutritional preferences confounds interpretations of giving-up density experiments." *Ecology* 99: 1517–1522.

Melange, J., P. Izar, and H. Japyassu. 2016. "Personality and behavioral syndrome in *Necromys lasiurus* (Rodentia: Cricetidae): Notes on dispersal and invasion process." *Acta Ethologica* 19: 189–195.

Montiglio, P. O., D. Garant, F. Pelletier, and D. Réale. 2012. "Personality differences are related to long-term stress reactivity in a population of wild eastern chipmunks, *Tamias striatus*." 84: 1071–1079.

Morris, D. W. 2005. "Habitat-dependent foraging in a classic predator-prey system: A fable from snowshoe hares." *Oikos* 109: 459–468.

Morris, G., L. M. Conner, and M. K. Oli. 2011. "Effects of mammalian predator exclusion and supplemental feeding on space use by hispid cotton rats." *Journal of Mammalogy* 92: 583–589.

O'Farrell, M. J. 1974. "Seasonal activity patterns of rodents in a sagebrush community." *Journal of Mammalogy* 55: 809–823.

Olsson, O., and J. S. Brown. 2010. "Smart, smarter, smartest: Foraging information states and coexistence." *Oikos* 119: 292–303.

Orrock, J. L., and R. J. Fletcher. 2014. "An island-wide predator manipulation reveals immediate and long-lasting matching of risk by prey." *Proceedings of the Royal Society B* 281: 20140391.

Pratas-Santiago, L. P., A. L. S. Gonçalves, A. M. V. Maia Soares, and W. R. Spironello. 2016. "The moon cycle effect on the activity of ocelots and their prey." *Journal of Zoology* 299: 275–283.

Réale, D., N. J. Dingemanse, A. J. N. Kazem, and J. Wright. 2010. "Evolutionary and ecological approaches to the study of personality." *Philosophical Transactions of the Royal Society B* 365: 3937–3946.

Réale, D., S. M. Reader, D. Sol, P. T. McDougal, and N. J. Dingemanse. 2007. "Integrating animal temperament within ecology and evolution." *Biological Review* 82: 291–318.

Roche, D. G., V. Careau, and S. A. Binning. 2016. "Demystifying animal 'personality' (or not): Why individual variation matters to experimental biologists." *Journal of Experimental Biology* 219: 3832–3843.

Rödel, H. G., M. Zapka, S. Talke, T. Kornatz, B. Bruchner, and C. Hedler. 2015. "Survival costs of fast exploration during juvenile life in a small mammal." *Behavioral Ecology and Sociobiology* 69: 205–217.

Scheibler, E., C. Roschlau, and D. Brodbeck. 2014. "Lunar and temperature effects on activity of free-living desert hamsters (*Phodopus roborovskii*, Satunin 1903)." *International Journal of Biometeorology* 58: 1769–1778.

Schmidt, K. A. 2000. "Interactions between food chemistry and predation risk in fox squirrels." *Ecology* 81: 2077–2085.

Schrader, A. M., B. P. Kotler, J. S. Brown, and G. I. H. Kerley. 2008. "Providing water for goats in arid landscapes: Effects on feeding effort with regard to time period, herd size and secondary compounds." *Oikos* 117: 466–472.

Schradin, C. 2006. "Whole-day follows of striped mice (*Rhabdomys pumilio*) a diurnal murid rodent." *Journal of Ethology* 24: 37–43.

Sih, A., A. M. Bell, J. C. Johnson, and R. E. Ziemba. 2004. "Behavioral syndromes: An integrative review." *Quarterly Review of Biology* 79: 241–277.

Smyth, B., and S. Nebel. 2013. "Passive integrated transponders (PIT) tags in the study of animal movement." *Nature Education Knowledge* 4: 3.

Sovie, A. R., D. U. Greene, C. F. Frock, A. D. Potash, and R. A. McCleery. 2019. "Ephemeral temporal partitioning may facilitate coexistence in competing species." *Animal Behaviour* 150: 87–96.

Suraci, J. P., M. Clinchy, B. Mugerwa, M. Delsey, D. W. Macdonald, J. A. Smith, et al. 2017. "A new Automated Behavioural Response system to integrate playback experiments into camera trap studies." *Methods in Ecology and Evolution* 8: 957–964.

Sutherland, D. R., and M. Predavec. 2010. "Universal trap timer design to examine temporal activity of wildlife." *Journal of Wildlife Management* 74: 906–909.

Van Der Merwe, M., and J. S. Brown. 2008. "Mapping the landscape of fear of the cape ground squirrel (*Xerus inauris*)." *Journal of Mammalogy* 89: 1162–1169.

Vickery, W. L., and J. R. Bider. 1981. "The influence of weather on rodent activity." *Journal of Mammalogy* 62: 140–145.

Walsh, R. N., and R. A. Cummins. 1976. "The open field test: A critical review." *Psychological Bulletin* 83: 481–504.

Wolff, J. O. 1993. "Why are female small mammals territorial?" *Oikos* 68: 364–370.

13. Predator-Mediated Sampling of Small Mammal Communities

Background

Predator diet studies were among the earliest studies conducted to understand predator-prey relationships (Errington 1930, Hamilton and Hunter 1939). These studies identified prey based on contents of the digestive tract (Hamilton and Hunter 1939, Pollack 1951) and remains in feces (Dearborn 1932) or raptor pellets—the regurgitated remains of less digestible components of prey such as bones, teeth, and fur (Errington 1930). It is no surprise that scientists soon considered using predator diets to gain insight into prey communities (Yom-Tov and Wool 1997). In some cases, using predator diets to census small mammals has distinct advantages.

Aims

In this chapter, we explore the strengths and weaknesses of using predator diets to derive inferences about small mammal populations and communities. We examine methodologies for using predator diets to determine the presence and distribution of small mammal species and the richness and composition of communities. Additionally, we look at how the choice of predators influences our inferences of small mammal populations and communities, and we look at the overall efficacy of using predators to study small mammals. We finish the chapter with a detailed description of the collection, processing, and analysis of predator diets.

Advantages

Live-trapping has long been the preferred method of sampling small mammals, and capture-recapture approaches are likely to remain the dominant approach for estimating small mammal abundance and population dynamics (Chapter 9). However, use of live traps is labor intensive and risks injury to captured animals (Matos et al. 2015); thus, justification for using live traps is often requested by institutional animal care and use committees (IACUCs). Most IACUCs adhere to the three Rs—replacement, reduction, and refinement (Russell and Burch 1959)—when considering approval of research proposals. In studies of free-ranging wildlife, replacement of study species is often not possible because the species itself may be the focus of the study. However, some study objectives can reduce or eliminate the number of animals handled by using indirect sampling methods (Chapter 2). Additionally, some small mammal species are difficult to identify without sacrificing the animal or taking tissue samples for later genetic analysis—a clear problem when dealing with species of conservation concern. For example, Torre et al. (2015) used barn owl (*Tyto alba*) pellets to assess abundance and spatial distribution of the yellow-necked mouse (*Apodemus flavicollis*). The yellow-necked mouse is almost impossible to distinguish from its more abundant congener, *A. sylvaticus*, without sacrificing the animal and examining molar shapes or using molecular techniques. Using barn owl pellets, the researchers estimated the abundance of yellow-necked mice relative to other species in the area using molar shape for species identification. Accordingly, predator-mediated collection of small mammal data can have some advantages over live-trapping (Table 13.1) and can provide acceptable indirect measures of small mammal

Table 13.1. Advantages and disadvantages of predator-mediated data collection to inventory small mammal populations relative to live trapping

Advantages	Disadvantages
Cost effective	Scats/pellets may be difficult to find
Less time-consuming	Difficult to link prey to vegetation types
Better richness estimates possible	Activity patterns of predator may affect relative proportion of prey detected
Works in areas where trapping is not feasible	Density estimates not possible
Better detection of rare species	Predator populations can vary, affecting long-term monitoring
May offer improved species identification	
Noninvasive sampling method	

presence, relative abundance, and diversity. Raptor pellets (Love et al. 2000, Meek et al. 2012), and to a lesser extent carnivore scats (Ray 1998, Torre et al. 2004, Price et al. 2005), are most often used when considering predator diet to inventory small mammal communities. Because snake diets are usually determined by collecting feces while handling captured snakes (Goetz et al. 2016) or by extracting stomach contents from snakes held in museum collections (Webber et al. 2016), using snake diets to census small mammals likely has limited application.

Limitations and General Assumptions

Using predator diets to study small mammals is an appealing idea, but there are clear limitations. Predators move over relatively large areas, meeting their nutritional needs by foraging in multiple patches and patch types (Deuel et al. 2017). Thus, while it is possible to link small mammals to landscape composition (Love et al. 2000, Lyman 2012), predator diets do not allow for the estimation of patch-specific small mammal population metrics. For similar reasons, using predator diets to study small mammal use of vegetation types is problematic. Where study objectives can be met using predator diets, there are assumptions that should be considered when interpreting resulting data; the difficulty in meeting these assumptions varies substantially depending upon study objectives.

Species Presence

For studies seeking to determine presence of a particular small mammal species, assumptions are not overly burdensome, most notably that the predator or predators used for diet assessment actually consume the small mammal and that the number of predator diet samples (e.g., number of pellets available for analyses) is sufficient to detect consumption if the small mammal is present. Here, presence is confirmed by any detection of the small mammal in the scat or pellet. In studies of this nature, the researcher is primarily concerned with sufficient sample sizes as to reduce probability of false negatives, i.e., the species was present but not detected. When planning new studies, a literature review of the range-wide diets of predators found within the study area may provide cues regarding feasibility of meeting this assumption.

Determining species presence based on predator diets logically extends to monitoring trends in small mammal populations over time or comparing their populations among areas. Making these comparisons is relatively straightforward assuming the predator consumes the small mammal of interest regularly and the predator exists through time or among areas. When monitoring trends over time or comparing small mammal populations among areas, an additional assumption is required—alternative prey are assumed to have equal impact on predator diets. For example, Lôpez-Bao et al. (2010) noted that Iberian lynx (*Lynx pardinus*) preyed almost exclusively on the European wild rabbits (*Oryctolagus cuniculus*). These lynx were supplementally fed domestic rabbits in an attempt to maintain lynx populations. Although food supplementation was provided to lynx populations, consumption of wild rabbits consistently tracked abundance of wild rabbits in nature, suggesting the Iberian lynx diet was a reliable source of data for monitoring European rabbit populations.

Species Richness

For studies attempting to estimate small mammal species richness, assumptions become more difficult. Here researchers must assume that sampling methods are capable of detecting presence of *all* small mammal species in the area of interest. This can be problematic, as comparative studies indicate fairly large sampling biases among traditional (e.g., live traps) sampling methods (Wiener and Smith 1972, Pearson and Ruggiero 2003, Umetsu et al. 2006), and similar biases occur when using predator diets to sample small mammal communities (Yom-Tov and Wool 1997, Torre et al. 2004, Avenant 2005, Rocha et al. 2011, Matos et al. 2015). Sampling intensity is also important when estimating small mammal species richness, and methods to ensure sufficient sampling intensity (discussed later in this chapter) should be employed. Thus, if monitoring species richness over time, sampling methods and effort (e.g., number of scats or pellets) must be relatively consistent over time, or analytical methods should be used to minimize effects of varying sampling intensity.

Community Composition

When the goal of a study is to understand not only species richness but small mammal community composition (i.e., the proportion of each species in the community; Chapter 10) assumptions may be onerous. Yom-Tov and Wool (1997) discussed two important assumptions associated with using barn owl pellets to estimate proportion of small mammals in the wild; these assumptions also hold for other forms of predator-mediated data collection. The first assumption is that prey consumed by the predator is a random sample of prey in the field (i.e., there is no predator selection among prey items). When using traps to measure community composition, a similar assumption—the trapping approach is a random sample of prey in the field—is also required. When using traps to sample small mammal species richness, abundance, and community structure, results can vary based on trap type (Wiener and Smith 1972, Umetsu et al. 2006) and trapping configuration (Pearson and Ruggiero 2003); thus, different trapping methods are very likely to estimate different community compositions. This makes assessment of the Yom-Tov and Wool's (1997) first assumption extremely difficult in natural settings because there is no "gold standard" for evaluating predator diets relative to prey availability in the wild. Thus, this first assumption can only be tested when the small mammal community composition is known, a condition that is rarely met.

The second assumption associated with using predator diets to estimate proportion of small mammals is that the pellets or scats themselves represent an unbiased sample of small mammals consumed. This latter assumption was assessed via computer simulations (Yom-Tov and Wool 1997). They concluded that consumption of larger prey resulted in more single-prey pellets than expected if pellets were a random sample of prey consumed, and they suggested that this should be taken into consideration when estimating the distribution of small mammals in the wild. Although this is a valid consideration when establishing a research or monitoring protocol, this concern may be reduced by careful selection of predator or predators for which diet data are collected (discussed later in this chapter).

Comparative Studies

Although a gold standard is lacking for comparing differing approaches, there have been a number of studies comparing predator diets and trapping approaches at assessing small mammal communities. These studies suggest that use of predator diet to quantify small mammal communities may be dependent on study area, choice of predator, or on the small mammal community itself.

Rosellini et al. (2008) took a unique approach to determining whether European marten (*Martes martes*) diet reflected seasonal abundances of small mammals. Although small mammal community composition was not known, small mammals were most abundant in fall and least abundant during winter and spring. Live-trapping mirrored seasonal fluctuations but European martens continued to select small mammals over other prey regardless of season, suggesting European marten were not opportunistic predators.

In another study, Torre et al. (2004) compared small mammal species-richness estimates based on data collected from scats of the common genet (*Genetta genetta*), barn owl pellets, and live-trapping. They also compared these estimates against the known number of small mammal species available in the study area. Barn owls were superior regarding detection of insectivores and grassland rodents, whereas woodland rodents were more readily detected in genet scats. Sample sizes varied greatly among methods with >17,000 individuals sampled using owl pellets, >2000 individuals sampled in genet scats, and >1400 individuals captured using live-trapping. They concluded that owl pellets provided higher richness estimates when the number of pellets sampled was <50, whereas richness estimates using genet scats were higher when >100 scats were sampled. Interestingly, there were 19 small mammal species present in the study area and richness estimates obtained from barn owls (17 species) and genets (14 species) were superior to estimates obtained by live-trapping (9 species). In contrast, Rocha et al. (2011) compared species-richness estimates derived from barn owl pellets to estimates obtained using live-trapping. Two species were unique only to barn owl pellets, but four species captured in live traps were not detected in pellets, leading the researchers to conclude that live-trapping provided greater species-richness estimates, but the methods were complimentary.

Activity patterns of predators and prey potentially affect small mammal occurrence in scats and pellets. For example, some small mammals are nocturnal, some crepuscular, some diurnal, and others active throughout the diel period. As a result, predator activity patterns should be taken into consideration when using scats or pellets to inventory small mammal communities. Similar to Torre et al. (2004), Matos et al. (2015) compared richness, diversity, evenness, abundance, and proportion of each species sampled using live-trapping, pitfall traps, and marsh harrier (*Circus aeruginosus*) pellets. Importantly, marsh harriers are entirely diurnal. In keeping with the results of Torre et al. (2004), they found predator-mediated data collection was superior to live-trapping and pitfalls with regard to both inventorying small mammals and cost-effectiveness. In their study, 14 total species were detected among all methods; 11 species were detected in harrier pellets, whereas live-trapping detected six species and pitfalls five. Importantly, they suggested that prey in marsh harrier pellets underestimated relative abundance of nocturnal small mammals and overestimated abundance of diurnal prey.

In contrast to both Torre et al. (2004) and Matos et al. (2015), Rocha et al. (2011) observed greater species richness associated with live-trapping than detected within barn owl pellets. There were four species detected in live traps that were not detected in owl pellets, whereas only two species were detected in owl pellets that were not detected in live traps. Interestingly, the species unique to owl pellets represented genera that were not previously known to the study area. They concluded that owl pellets were a valuable resource and should be included with live-trapping efforts when surveying small mammals.

Despite the assertion that barn owls forage randomly and thus consume prey as available on the landscape, Andrade et al. (2016) and empirical evidence presented above (Torre et al. 2004, Rocha et al. 2011, and Matos et al.

2015) suggest otherwise (i.e., in all studies other techniques detected species that were not detected in barn owl diets). Thus, great care should be taken when interpreting data with regard to small mammal community composition. However, comparing small mammal communities among different places or across time using the diet of a common predator (i.e., same predator diet for all areas or times) may be worthwhile. For example, Clark and Bunck (1991) used published studies of barn owl diets spanning approximately 70 years to evaluate environmental factors affecting population trends in North American small mammals. Because they focused on the diet of a single predator species, it was reasonable to assume that differences in diet reflected differences in prey communities. Similarly, Avery et al. (2005) examined owl pellets from 64 localities in the Western Cape province of South Africa, recording 56 species of small mammals. Apart from providing numerous examples of species range extensions, they were also able to infer ecologically important patterns such as seasonality of breeding.

Choice of Predator Species

Retroactive examination of predator studies that concurrently sampled prey abundance (for example, see Baker et al. 2001) may provide insight regarding potential efficacy of using a particular predator's diet for studying small mammals. If relevant studies do not exist in the literature and preliminary research cannot be performed, some general guidelines regarding usefulness of potential predators for studying small mammals are available as a starting point.

Predator diet studies are generally conducted to determine the importance of given prey to a predator. As a result, the choice of predator studied is dictated by the study objective. In other words, which predator diet is to be studied? In contrast, choice of predator or predators to use when studying small mammal species or communities should be based on the likelihood of sampling the desired parameters of the small mammal study under consideration and the likelihood of obtaining a sufficient sample (i.e., a sufficient number of scats, pellets, or stomach/intestinal contents) such that resulting estimates have the desired level of accuracy and precision while meeting the assumptions associated with overall study objectives.

One criterion for selecting a predator to census small mammals is the study area itself. Obviously, the predator must be present on the study area to be useful, but there are other—perhaps less obvious—considerations as well. Because predators have large home ranges relative to their prey, predator diets should be considered as aggregators of small mammal occurrence over areas much larger than the home range of the small mammals being studied. As a result, predator diets should only be used to make inferences regarding small mammals on relatively large study areas. Examination of studies using owl pellets to make inferences dealing with small mammals suggest this is understood, if not explicitly discussed—for example, a 700 km^2 natural park (Torre et al. 2015) or the British countryside (Love et al. 2000).

Given an adequately large study area, primary criteria for selecting predators for sampling small mammals include (1) the predator is sufficiently abundant in the study area and uses the area for which small mammal information is required, (2) the predators' scats or pellets can be correctly identified and located in sufficient numbers to provide confidence in resulting estimates, or in the case of stomach contents, there is a readily available source of stomach and intestinal contents for the predator, and (3) the predator is opportunistic, consuming all small mammals of interest. When research is focused on a single species of prey, this third criteria can be relaxed such that the predator simply preys upon or even specializes in preying upon the small mammal of interest. In this latter case, sampling across time can provide trend data for the small mammal under study (Love et al. 2000, Lyman 2012).

In general, all of these criteria are most easily met in midsized mammalian carnivores and small and midsized birds of prey. Although small mammals are important in the diet of many snakes, the difficulty of obtaining sufficient sample sizes may restrict use of snake diets to quantifying historical small mammal communities based on diets derived from snakes held in museum collections (Glaudas et al. 2017). Among mammals, small felids perhaps warrant the most consideration regarding use of their diet to census small mammals. Small (≤15 kg) (Wallach et al. 2015) felids are found in most of the world's terrestrial ecosystems and are almost entirely carnivorous; small mammals can be expected to serve as important prey for them. Additionally, small wild felids often exist in sufficient numbers as to make their diets worthy of consideration for assessing small mammal communities (Ramesh and Downs 2015). Finally, felid scats can generally be distinguished from scats of nonfelid carnivorous species; however, scats can be difficult to distinguish among similarly sized felids (Davison et al. 2002).

Among birds of prey, small and midsized owls are most promising because their pellets generally contain significant remains of prey consumed, making identification of prey consumed relatively straightforward (Marti et al. 2007). Although Matos et al. (2015) used marsh harriers to assess species richness, they also noted that pellets were difficult to locate. Additionally, falconiform (i.e., diurnal birds of prey) pellets often do not contain significant remains of prey consumed (Marti et al. 2007), making them less desirable for evaluating small mammal occurrence. Larger owls often consume larger prey, which require "dismantling" prior to consumption; thus, these owls may also prove less valuable for determining small mammal abundance.

Existing studies using carnivore scats and raptor pellets to quantify aspects of prey species populations and communities are overwhelmingly dominated by studies using barn owls (Yom-Tov and Wool 1997). There are a number of factors contributing to the popularity of using barn owl pellets to make inferences regarding small mammal communities. Barn owls are near cosmopolitan in their distribution (Yom-Tov and Wool 1997, Rocha et al. 2011), thus their pellets are available for study to many researchers. They also have very strong fidelity to roost sites, which facilitates collection of pellets (Yom-Tov and Wool 1997, Avery et al. 2005). On average, barn owls regurgitate an average of 1.4 pellets per day (Yom-Tov and Wool 1997), a number capable of producing acceptable sample sizes from only a few roosts. Finally, because barn owls swallow most prey whole, pellets contain sufficient skeletal elements (Fig. 13.1) to facilitate relatively straightforward identification of prey (Andrews 1990). Because barn owls are the most-used species for censusing small mammals, there are many examples of their use for studying various aspects of small mammal populations and communities.

Hodara and Poggio (2016) studied changes in barn owl diets relative to loss of habitat resulting from agricultural intensification. Barn owls foraged more on small mammals before intensification of agriculture than after. Hodara and Poggio suggested continued monitoring of barn owl diets as a mechanism for monitoring change in the faunal community. Similarly, Lyman (2012) used barn owl pellets to detect a change in small mammal fauna resulting from conversion of agriculture to grassland, demonstrating pellets were capable of detecting differences over short time spans. Kross et al. (2016) used barn owl pellets to document decline in certain prey species associated with increased acreage devoted to perennial agriculture. Love et al. (2000) also demonstrated that barn owl pellets indicated how small mammals responded to land-use change in Great Britain. As land was taken out of agriculture, wood mice, yellow-necked mice, bank voles (*Clethrionomys glareolus*), and pygmy shrews (*Sorex minutus*) increased in barn owl pellets and common shrews (*S. araneus*) decreased.

Owl diets are not restricted to nonvolant mammals. Khalafalla and Ludica (2012) documented big brown bats (*Eptesicus fuscus*) as prey of barn owls, and Love et al. (2000) found remains of bats in barn owl pellets. Owl foraging on bats was observed by one of the authors (LMC) when he observed an eastern screech owl (*Megascops asio*) repeatedly capturing southeastern myotis (*Myotis austroriparius*) as they emerged from a cave in southwestern Georgia, USA.

Fig. 13.1. Intact barn owl pellet prior to dissection (A) and sorted remains from barn owl pellets (B). Notice relatively intact skulls and mandibles, which facilitate accurate identification of small mammal species.

Finally, although the use of mammalian carnivore diets to survey small mammal populations is uncommon relative to use of raptor pellets, the scientific literature contains a plethora of carnivore diet studies (Beasom and Moore 1977, Miller and Speake 1978, Maehr and Brady 1986, Godbois et al. 2003). Because of this, existing carnivore diet studies may offer insight into changes in small mammal communities that may occur over long time scales.

Sample Collection

For those predators that frequently consume small mammal prey, stomach-content analysis often involves lethal sampling (Beasom and Moore 1977, Fritts and Sealander 1978, Maehr and Brady 1986, Litvaitis et al. 1996, McDonald et al. 1996). In general, samples are obtained from predators that were harvested for fur (Fritts and Sealander 1978), for sport, or as part of predator control programs (Lang 2008). Care should be taken when interpreting stomach and intestinal contents from predators that were obtained by trapping; samples from these animals may contain bait ingested at the trapping set, and actual prey may have been passed while the predator was in the trap. Animals obtained by hunting or by kill traps, such as body gripping traps and snares, are generally considered to provide less biased data regarding their diet (Litvaitis et al. 1996, McDonald et al. 1996). In most cases, animals obtained from harvest by hunters and trappers will not likely represent year-round diets, as hunting and trapping seasons are often limited to a few months of the year (Litvaitis et al. 1996); thus, conclusions regarding small mammal populations or communities should consider potential seasonality effects when drawing inferences. Collection of stomach contents from snakes preserved as museum specimens (R. W. Clark 2002, Hoyos and Almeida-Santos 2016, Webber et al. 2016) may provide data useful for addressing some study objectives. However, museum specimens often lack specific collection protocols; thus, use of such data to make inferences should be made with care (Glaudas et al. 2017).

Scats from mesomammal predators have seldom been used to quantify small mammal communities (but see Torre et al. 2004 and Rosellini et al. 2008). However, as discussed above, some mesopredators may be suitable for studying small mammal communities. When initiating a study using meso-predator diet to indirectly quantify small mammals, scat collection protocols associated with traditional predator diet studies (Godbois et al. 2003, Gulsby et al. 2015, Cherry et al. 2016) may prove beneficial. Being able to identify predator scats of interest from those of other sympatric predators is important, and initial training from someone familiar with identifying spoor of the target species can be invaluable (Prugh and Ritland 2005). However, even experienced observers can misidentify scat (Morin et al. 2016). One way to be sure of the identity of the predator producing the scat is to use telemetry by, for example, fixing a GPS-tracker on the predator (Leighton et al. 2020).

Morin et al. (2016) used genetic analyses to determine observer ability to identify bobcat, coyote, and black bear scats. They found that bobcat and coyote scats were frequently confused and suggested genetic sampling to ensure correct identification of predators. We suggest field personnel should initially be mentored by someone familiar with predator scats of the area, and mitochondrial sequencing (Gulsby et al. 2016) should be performed early in the study to ensure accuracy of scat identification. If identification is deemed unacceptable, it may be necessary to use molecular approaches to ensure correct predator identification for the duration of the study.

When predator diet is estimated from predator scats, field collection of scats generally occurs systematically (Godbois et al. 2003) or opportunistically (Cherry et al. 2016). We suggest systematic collection of scats from along roads and trails offers the greatest advantage when the goal is to census small mammals. Systematic sampling ensures a consistent collection effort over time and space, thus increasing the ability to detect differences among sites or trends over time. Collection along roads and trails also increases the ability to detect scats and reduces the likelihood that scats may be overlooked. Use of scat detection dogs (Wasser et al. 2004) may further increase scat collection rates and efficiently increase the area sampled.

Care should be taken to perform scat collections with sufficient frequency to facilitate identification of the scat. When species depositing the scat cannot be determined, scats should be excluded from analyses. Importantly, prey size can affect persistence of scats, thus affecting ability to identify the predator responsible for depositing the scat. This can lead to biased estimates of consumption (Godbois et al. 2005). As a result, frequent scat collection intervals (<4 weeks) are preferred. More frequent collection should be considered in harsh climates.

Raptor pellet collection can be facilitated by locating roost sites (Fig. 13.2). Barn owl roosts in particular are often use for decades (Yom-Tov and Wool 1997). Barn owl propensity to roost in protected areas, such as barns, silos, sheds, and caves, results in persistence of pellets due to lack of exposure to the elements. As a result, pellet accumulation in these areas can be great. In these cases, researchers will benefit from an initial collection of pellets followed by periodic collections based on the time intervals of interest.

Sample Processing

Methods for assessing predator diets can be classified into two major categories, those involving stomach- or intestinal-content analyses (Beasom and Moore 1977, Fritts and Sealander 1978, Litvaitis et al. 1996, McDonald et al. 1996) and those based on analyses of scats (Klare et al. 2011) or pellets (Salamolard et al. 2000). In both cases, identification of prey remains from hair and bone is required, though actual processing procedures may vary.

Fig. 13.2. The barn owl's fidelity for roost sites often allows researchers to easily obtain sufficient pellet samples. (*A*) Pellets underneath a roost prior to initial collection. Notice hair and bone from pellets that decomposed prior to the roost being discovered by researchers. (*B*) Pellets underneath a barn owl roost that is being regularly visited by researchers. These pellets can provide temporal data regarding small mammal communities because the previous collection date bounds the timeframe available for pellets to accumulate.

Stomach-Content Analyses

When processing stomach and intestinal samples, the samples should be washed with hot water through a large sieve, and prey remains should be identified using reference collections of small mammal bones and hair. Latex gloves, eye protection, and a surgical mask should be used when handling stomach and intestinal contents, as many predators harbor diseases and parasites that are transmissible to humans.

Fecal- or Pellet-Content Analyses

Analyses of fecal and pellet contents are the most common methods used to study diets of terrestrial vertebrate predators (mammals, Baker et al. 2001,

Godbois et al. 2003, Cherry et al. 2016; raptors, Marti et al. 2007, Escribano et al. 2016; and snakes, Mociño-Deloya et al. 2015, Goetz et al. 2016). Methods of handling scats and pellets for diet analyses vary based on individual researcher preferences. Some researchers oven-dry and then dissect scats or pellets to identify prey consumed (Godbois et al. 2003, Cherry et al. 2016, Escribano et al. 2016), whereas others wash remains to remove the fecal or pellet matrix before analyses of skeletal remains (Love et al. 2000, Klare et al. 2011). Marti et al. (2007) suggest that soaking and washing pellets with water or a 10% solution of sodium hydroxide is preferred when sample sizes are large or where better diet resolution is required. As with stomach-content analyses, simple protective gear should be used to protect researchers from potentially harmful agents present in scats or pellets. Simple identification of prey is often sufficient to address some objectives regarding small mammal populations or communities (e.g., species presence or species richness). However, additional efforts may be required if trying to estimate relative abundance of different small mammals consumed.

When estimating relative abundance of small mammals consumed, it is required that the number of small mammals consumed be estimated. Data from owl pellets, particularly barn owl pellets, seldom require special treatment to estimate number of individuals consumed as skeletal remains are often sufficiently complete as to represent counts of prey consumed. However, this is not the case when using carnivore scats. Klare et al. (2011) reviewed techniques for estimating diets of terrestrial carnivores based on scat analyses. They concluded that, although uncommon, the best estimates of predator diets were obtained when biomass calculation models were applied to the data. Biomass models are developed using known carnivore diets and are designed to estimate biomass consumed by the predator from some quantitative estimate of remains in scat. When biomass calculation models do not exist, it may be possible to use models based on feeding trials of closely related species.

Molecular Analyses

Molecular analyses of predator scats have been used to identify both predator species and their prey (Farrell et al. 2000). Mumma et al. (2016) observed molecular analyses of scats to provide results similar to morphological approaches with regard to ranking of importance of prey in the diet, but molecular analyses detected prey more frequently. However, they also point out that proper homogenization—ensuring that scat samples are thoroughly mixed such that prey DNA is evenly dispersed within the sample—is important to avoid false negatives using molecular approaches. It may be possible to homogenize all scat samples to address some study objectives (e.g., species presence), making molecular approaches particularly cost-effective. In a study to quantify leporid consumption by coyotes, molecular approaches detected prey more frequently than morphological analyses, but location of the sample taken within the scat affected detection, further emphasizing the importance of sample homogenization (Gosselin et al. 2017).

Sample Size Assessment

When using pellets or scats to survey small mammal populations and communities, species-area curves can be used to determine when adequate sample sizes have been attained. This said, sampling pellets until 150 individual prey items were identified seems to be used as a rule of thumb regarding adequacy of sample size (van Strien et al. 2015). Another method for deriving better richness estimates involves establishing detection probabilities using occupancy analysis (MacKenzie et al. 2006), as demonstrated by van Strien et al. (2015), who advocated splitting batches of pellets into two equal samples and using these to estimate probability of detection as though they represented temporal replicates, and who concluded the approach produced trend estimates with sufficiently small confidence intervals.

Occasionally, existing data may be used to make inferences regarding small mammal populations or communities. When these data are used to make comparisons among sites or over time, researchers must ensure adequacy of samples to ensure comparisons are not influenced by unequal sample sizes among sites or times being compared. In the event sample sizes are entirely adequate based on species-area curves, as described above, statistical inference can be straightforward. However, if sample sizes differ among study sites or among time intervals, researchers face a dilemma—should random samples from one place or time be deleted to place data sets on equal footing or should statistical procedures be used to estimate and correct for lack of samples within the smaller data set? Deleting samples is straightforward, but can result in a loss of information. In contrast, explicitly accounting for differences in sampling effort (Torre et al. 2004) can be used to compare data sets that differ regarding sample sizes.

Conclusions and Future Directions

Using predator diets to study small mammals presents a noninvasive approach to addressing some ecological questions regarding species distribution, species richness, and community composition. Because it is nearly impossible to evaluate small mammal sampling methods regarding their ability to accurately reflect community composition in natural systems, comparisons among sampling methods are best used to provide insight into the suitability of a given method for sampling a given species or group of species. In some cases, however, a thorough review of literature associated with particular predators and the prey they consume may provide the necessary information for establishing study protocols.

Care should be exercised when using predator diets to infer relative abundance on small mammal species in the area of interest, as predator diet preferences can affect study results. However, studies comparing small mammal communities among two or more areas or over time may obtain useful results in spite of predator dietary preferences if the predator monitored remains consistent.

It is likely that determining small mammal species richness for an area would benefit from combining data from several sources to include traditional trapping approaches and data collected based on predator diet analyses (Torre et al. 2004). Likewise, this same recommendation holds when seeking to document occurrence of a particular small mammal in a given area.

Future conservation efforts may benefit from literature reviews and meta-analyses that compare historic predator diets to current prey species distributions (e.g., Avery 1987). Such approaches offer promise as means of evaluating changes in small mammal distributions relative to long-term environmental change (e.g., climate change, land-use intensification).

REFERENCES

Andrade, A., J. F. S. de Menezes, and A. Monjeau. 2016. "Are owl pellets good estimators of prey abundance?" *Journal of King Saud University—Science* 28: 139–244.

Andrews, P. 1990. *Owls, Caves and Fossils.* Chicago: University of Chicago Press.

Avenant, N. L. 2005. "Barn owl pellets: A useful tool for monitoring small mammal communities?" *Belgian Journal of Zoology* 135 (supplement): 39–43.

Avery, D. M. 1987. "Late Pleistocene coastal environment of the Southern Cape Province of South Africa: Micromammals from klasies river mouth." *Journal of Archaeological Science* 14: 405–421.

Avery, D. M., G. Avery, and N. G. Palmer. 2005. "Micromammalian distribution and abundance in the Western Cape Province, South Africa, as evidenced by barn owls *Tyto alba* (Scopoli)." *Journal of Natural History* 39 (22): 2047–2071.

Baker, L. A., R. J. Warren, D. R. Defenbach, W. E. James, and M. J. Conroy. 2001. "Prey selection by reintroduced bobcats (*Lynx rufus*) on Cumberland Island, Georgia." *The American Midland Naturalist* 145: 80–93.

Bartlett, R., and P. Bartlett. 2005. *Guide and Reference to the Snakes of Eastern and Central North America (North of Mexico).* Gainesville: University Press of Florida.

Beasom, S. L., and R. A. Moore. 1977. "Bobcat food habit response to a change in prey abundance." *Southwestern Naturalist* 21: 451–457.

Cherry, M. J., K. L. Turner, M. B. Howze, B. S. Cohen, L. M. Conner, and R. J. Warren. 2016. "Coyote diets in a longleaf pine ecosystem." *Wildlife Biology* 22: 64–70.

Clark, D. R., and C. M. Bunck. 1991. "Trends in North American small mammals found in common barn owl (*Tyto alba*) dietary studies." *Canadian Journal of Zoology* 69: 3093–3102.

Clark, R. W. 2002. "Diet of the timber rattlesnake (*Crotalus horridus*)." *Journal of Herpetology* 36: 494–499.

Crooks, K. R., and M. E. Soule. 1999. "Mesopredator release and avifaunal extinctions in a fragmented system." *Nature* 400: 563–566.

Davison, A., J. D. S. Birks, R. C. Brookes, T. C. Braithwaite, and J. E. Messenger. 2002. "On the origin of faeces: Morphological versus molecular methods for surveying rare carnivores from their scats." *Journal of Zoology* 257: 141–143.

Dearborn, N. 1932. "Foods of some predatory furbearing animals in Michigan." *University of Michigan, School of Forestry and Conservation Bulletin* 1: 28–30.

Deuel, N. R., L. M. Conner, K. V. Miller, M. J. Chamberlain, M. J. Cherry, and L. V. Tannenbaum. 2017. "Habitat selection and diurnal refugia of gray foxes in southwestern Georgia, USA." *PLoS ONE* 12 (10): e0186402. https://doi.org/10.1371/journal.pone.0186402.

Errington, P. L. 1930. "The pellet analysis method of raptor food habits study." *Condor* 32: 292–296.

Escribano, N., D. Galicia, A. H. Ariño, and C. Escala. 2016. "Long-term data set of small mammals from owl pellets in the Atlantic-Mediterranean transition area." *Scientific Data* 3: 160085. https://doi.org/10.1038/sdata.2016.85.

Estes, J. A., J. Terbough, J. S. Brashares, M. E. Power, J. Berger, W. J. Bond, et al. "Trophic downgrading of planet Earth." *Science* 333: 301–306.

Farrell, L. E., J. Roman, and M. E. Sunquist. 2000. "Dietary separation of sympatric carnivores identified by molecular analysis of scat." *Molecular Ecology* 9: 1583–1590.

Fauvelle, C., R. Diepstraten, and T. Jessen. 2017. "A meta-analysis of home range studies in context of trophic levels: Implications for policy-based conservation." *PLoS ONE* 12: e017336.

Fredriani, J. M., T. K. Fuller, R. M. Sauvajot, and E. C. York. 2000. "Competition and intraguild predation among three sympatric carnivores." *Oecologia* 125: 258–270.

Fritts, S. H., and J. A. Sealander. 1978. "Diets of bobcats in Arkansas with special reference to age and sex differences." *Journal of Wildlife Management* 42: 533–539.

Fritzell, E., and K. Haroldson. 1982. "*Urocyon cinereoargenteus*." *Mammalian Species* 189: 1–8.

Glaudas, X., T. C. Kearney, and G. J. Alexander. 2017. "Museum specimens bias measures of snake diet: a case study using the ambush-foraging puff adder (*Bitis arietans*)." *Herpetologica* 73 (2): 121–128.

Godbois, I. A., L. M. Conner, B. D. Leopold, and R. J. Warren. 2005. "Effect of diet on mass loss of bobcat scat after exposure to field conditions." *Wildlife Society Bulletin* 33: 149–153.

Godbois, I. A., L. M. Conner, and R. J. Warren. 2003. "Bobcat diet on an area managed for northern bobwhite." *Proceedings of the Southeastern Association of Fish and Wildlife Agencies* 57: 222–227.

Goetz, S. M., C. E. Petersen, R. K. Rose, D. F. Kleopfer, and A. H. Savitzky. 2016. "Diet and foraging behaviors of timber rattlesnakes, *Crotalus horridus*, in eastern Virginia." *Journal of Herpetology* 50: 520–526.

Gosselin, E. N., R. C. Lonsinger, and L. P. Waits. 2017. "Comparing morphological and molecular diet analyses and fecal DNA sampling protocols for a terrestrial carnivore." *Wildlife Society Bulletin* 41: 362–369.

Gulsby, W. D., C. H. Killmaster, J. W. Bowers, J. D. Kelly, B. N. Sacks, M. J. Statham, and K. V. Miller. 2015. "White-tailed deer fawn recruitment before and after experimental coyote removals in central Georgia." *Wildlife Society Bulletin* 39: 248–255. https://doi.org/10.1002/wsb.534.

Gulsby, W. D., C. H. Killmaster, J. W. Bowers, J. S. Laufenberg, B. N. Sacks, M. J. Statham, and K. V. Miller. 2016. "Efficacy and precision of fecal genotyping to estimate coyote abundance." *Wildlife Society Bulletin* 40: 792–799.

Hamilton, W. J., and R. P. Hunter. 1939. "Fall and winter food habits of Vermont bobcats." *Journal of Wildlife Management* 3: 99–103.

Hodara, K., and S. L. Poggio. 2016. "Frogs taste nice when there are few mice: Do dietary shifts in barn owls result from rapid farming intensification?" *Agricultural Ecosystems and Environment* 230: 42–46.

Hoyos, M. A., and S. M. Almeida-Santos. 2016. "The South-American rattlesnake *Crotalus durissus*: Feeding ecology in the central region of Brazil." *Biota Neotropica* https://doi.org/10.1590/1676-0611-BN-2014-0027.

Kelly, B. T., and E. O. Garton. 1997. "Effects of prey size, meal size, meal composition, and daily frequency of feeding on the recovery or rodent remains from carnivore scats." *Canadian Journal of Zoology* 75: 1811–1817.

Khalafalla, S. M., and C. A. Ludica. 2012. "Barn owl (*Tyto alba*) predation on big brown bats (*Eptesicus fuscus*) in Pennsylvania." *Canadian Field Naturalist* 126: 38–40.

Klare, U., J. F. Kamler, and D. W. Macdonald. 2011. "A comparison and critique of different scat-analysis methods for determining carnivore diet." *Mammal Review* 41: 294–312.

Kross, S. M., R. P. Barbour, and B. L. Martinico. 2016. "Agricultural land use, barn owl diet, and vertebrate pest control implications." *Agriculture, Ecosystems, and Environment* 226: 167–174.

Lang, M. 2008. "The effects of intensive predator harvest during the quail nesting season on diet, age, and reproduction of meso-mammalian predators." Master's thesis, The University of Georgia, 61 pages.

Leighton, G. R. M., J. M. Bishop, M. J. O'Riain, J. Broadfield, J. Merondun, G. Avery, et al. 2020. "An integrated dietary assessment increases feeding event detection in an urban carnivore." *Urban Ecosystems* 23: 569–583.

Levi, T., and C. C. Wilmers. 2012. "Wolves-coyotes-foxes: A cascade among carnivores." *Ecology* 93: 921–929.

Litvaitis, J. A., K. Titus, and E. M. Anderson. 1996. "Measuring vertebrate use of terrestrial habitat and foods." In *Research and Management Techniques for Wildlife and Habitats*, edited by T. A. Bookhout, 254–274. Bethesda: The Wildlife Society.

López-Bao, J. V., A. Rodriguez, and F. Palomares. 2010. "Abundance of wild prey modulates consumption of supplementary food in the Iberian lynx." *Biological Conservation* 143: 1245–1249.

Love, R., C. Webbon, D. E. Glue, and S. Harris. 2000. "Changes in the food of British barn owls (*Tyto alba*) between 1975 and 1997." *Mammal Review* 30: 107–129.

Lyman, R. L. 2012. "Rodent-prey content in long-term samples of barn owl (*Tyto alba*) pellets from the northwestern United States reflects local agricultural change." *American Midland Naturalist* 176: 150–163.

MacKenzie, D. I., J. D. Nichols, J. A. Royle, K. H. Pollock, L. A. Bailey, and J. E. Hines. 2006. *Occupancy Modeling and Estimation*. San Diego: Academic Press.

MacWhirter, R., and K. Bildstein. 1996. "Northern harrier (*Circus cyaneus*)." In *The Birds of North America*, edited by A. Poole and F. Gill, species account no. 210. Washington, DC, and Philadelphia, PA: The Academy of Natural Sciences and The American Ornithologists' Union.

Maehr, D. S., and J. R. Brady. 1986. "Food habits of bobcats in southern Florida." *Journal of Mammalogy* 67: 133–138.

Marti, C. D., M. Bechard, and F. M. Jacksic. 2007. "Food habits." In *Raptor Research and Management Techniques*, edited by D. M. Bird, K. L. Bildstein, D. R. Barber, and A. Zimmerman, 129–153. Surrey: Hancock House Publishers Ltd.

Matos, M., M. Alves, M. J. Ramos Pereira, I. Torres, S. Marques, and C. Fonseca. 2015. "Clear as daylight: Analysis of diurnal raptor pellets for small mammal studies." *Animal Biodiversity and Conservation* 38: 37–48.

Mayer, M. V. 1952. "The hair of California mammals with keys to the dorsal guard hairs of California mammals." *American Midland Naturalist* 48: 480–512.

McDonald, L. L., J. R. Alldredge, M. S. Boyce, and W. P. Erickson. 1996. "Measuring availability and use of terrestrial habitats and foods." In *Techniques for Wildlife Investigations and Management*, edited by C. E. Braun, 465–488. Bethesda: The Wildlife Society.

Meek, W. R., P. J. Burman, T. H. Sparks, M. Nowakowski, and N. J. Burman. 2012. "The use of barn owl *Tyto alba* pellets to assess population change in small mammals." *Bird Study* 59: 166–174.

Merritt, J. F. 2010. *The Biology of Small Mammals*. Baltimore: Johns Hopkins University Press. 1–336.

Miller, S. D., and D. W. Speake. 1978. "Prey utilization by bobcats on quail plantations in southern Alabama." *Proceedings of the Annual Conference of the Southeastern Association of Fish and Wildlife Agencies* 32: 100–111.

Mociño-Deloya, E., K. Setser, M. Heacker, and S. Peurach. 2015. "Diet of New Mexico ridge-nosed rattlesnake (*Crotalus willardi obscurus*) in the Sierra San Luis and Sierra Pan Duro, Mexico." *Journal of Herpetology* 48: 104-107.

Morin, D. J., S. D. Higdon, J. L. Holub, D. M. Montague, M. L. Fies, L. P. Waits, and M. J. Kelly. 2016. "Bias in carnivore diet analysis resulting from misclassification of predator scats based on field identification." *Wildlife Society Bulletin* 40: 669–677.

Mumma, M. A., J. R. Adams, C. Zeminski, T. K. Fuller, S. P. Mahoney, and L. P. Waits. 2016. "A comparison of morphological and molecular analyses of predator scats." *Journal of Mammalogy* 97: 112–120.

Newsome, T. M., and W. J. Ripple. 2015. "A continental scale trophic cascade from wolves through coyotes to foxes." *Journal of Animal Ecology* 84: 49–59.

Pearson, D. E., and L. F. Ruggiero. 2003. "Transect versus grid trapping arrangements for sampling small-mammal communities." *Wildlife Society Bulletin* 31 (2): 454–459.

Pollack, M. E. 1951. "Food habits of bobcats in the New England states." *Journal of Wildlife Management* 15: 209–213.

Preston, C., and R. Beane. 1993. "Red-tailed hawk (*Buteo jamaicensis*)." In *The Birds of North America*, edited by A. Poole and F. Gill, species account no. 52. Washington,

DC, and Philadelphia, PA: The Academy of Natural Sciences and The American Ornithologists' Union.

Price, M. H. H., C. T. Darimont, N. N. Winchester, and P. C. Paquet. 2005. "Facts from faeces: Prey remains in wolf, *Canis lupus*, faeces revise occurrence records for mammals of British Columbia's coastal archipelago." *The Canadian Field Naturalist* 119: 192–196.

Prugh, L. R., and C. E. Ritland. 2005. "Molecular testing of observer identification of carnivore feces in the field." *Wildlife Society Bulletin* 33: 189–194.

Ramesh, T., and C. T. Downs. 2015. "Diet of serval (*Leptailurus serval*) on farmlands in the Drakensberg Midlands, South Africa." *Mammalia* 79 (4): 399–407.

Ray, J. C. 1998. "Temporal variation of predation on rodents and shrews by small African forest carnivores." *Journal of Zoology, London* 244: 363–370.

Rocha, R. G., E. Ferreira, Y. L. R. Leite, C. Fonseca, and L. P. Costa. 2011. "Small mammals in the diet of barn owls, *Tyto alba* (Aves: Strigiformes) along the mid-Araguaia River in central Brazil." *Zoologia* 28: 709–716.

Rosellini, S., I. Barja, and A. Pineiro. 2008. "The response of European pine marten (*Martes martes* L.) feeding to the changes of small mammal abundance." *Polish Journal of Ecology* 56: 497–503.

Russell, W. M. S., and R. L. Burch. 1959. *The Principles of Humane Experimental Technique*. London: Methuen.

Salamolard, M., A. Butet, A. Leroux, and V. Bretagnolle. 2000. "Responses of an avian predator to variations in prey density at a temperate latitude." *Ecology* 81: 2428–2441.

Smith, D. W., T. D. Drummer, K. M. Murphy, D. S. Guernsey, and S. B. Evans. 2004. "Winter prey selection and estimation of wolf kill rates in Yellowstone National Park, 1995–2000." *Journal of Wildlife Management* 68: 153–166.

Torre, I., A. Arrizabalaga, and C. Flaquer. 2004. "Three methods for assessing richness and composition of small mammal communities." *Journal of Mammalogy* 85: 524–530.

Torre, I., L. Fernández, and A. Arrizabalaga. 2015. "Using barn owl *Tyto alba* pellet analyses to monitor the distribution patterns of the yellow-necked mouse (*Apodemus flavicollis* Melchoir 1834) in a transitional Mediterranean mountain." *Mammal Study* 40: 133–142.

Tumlison, R. 1983. "An annotated key to the dorsal guard hairs of Arkansas game mammals and furbearers." *Southwest Naturalist* 28: 315–323.

Umetsu, F., L. Naxara, and R. Pardini. 2006. "Evaluating the efficiency of pitfall traps for sampling small mammals in the Neotropics." *Journal of Mammalogy* 87: 757–765.

van Strien, A. J., D. L. Bekker, M. J. J. La Haye, and T. can der Meij. 2015. "Trends in small mammals derived from owl pellet data using occupancy modeling." *Mammalian Biology* 80: 340–346.

Wallach, A. D., I. Izhaki, J. D. Toms, W. J. Ripple, and U. Shanas. 2015. "What is an apex predator?" *Oikos* 124: 1453–1461.

Wasser, S. K., B. Davenport, E. R. Ramage, K. E. Hunt, M. Parker, C. Clarke, and G. Stenhouse. 2004. "Scat detection dogs in wildlife research and management: Application to grizzly and black bears in the Yellowstone Ecosystem, Alberta, Canada." *Canadian Journal of Zoology* 82: 475–492.

Webber, M. W., T. Jezkova, and J. A. Rodriguez-Robles. 2016. "Feeding ecology of the sidewinder rattlesnake, *Crotalus cerastes* (Viperidae)." *Herpetologica* 72: 324–330.

Wiener, J., and M. Smith. 1972. "Relative efficiencies of four small mammal traps." *Journal of Mammalogy* 53: 868–873.

Wolf, C., and W. J. Ripple. 2017. "Range contractions of the world's large carnivores." *Royal Society Open Science* 4: 170052.

Yom-Tov, Y., and D. Wool. 1997. "Do the contents of barn owl pellets accurately represent the proportion of prey species in the field?" *The Condor* 99: 972–976.

14. Ecological and Taxonomic Study of Morphology

Background

Morphology in biology is the study of the size, shape, and structure of animals, plants, and microorganisms. This includes morphometrics—the quantitative analysis of form, encompassing both size and shape. In biology, morphology is used to test hypotheses concerning possible factors that influence size, shape, and structure in organisms, whether ontogenetic, phylogenetic, ecological, behavioral, or physiological. Early anatomical descriptions in the eighteenth and nineteenth centuries transitioned into more empirical and quantitative approaches in the twentieth century, culminating in the popularity of multivariate statistics ("traditional morphometrics") and phenetic approaches to taxonomy ("numerical taxonomy") in the 1960s to 1980s. Since the early 1990s, the advent of advanced statistical approaches to the geometric estimation of shape (Kendall et al. 1999) and the Procrustes superimposition method have led to the "geometric morphometric revolution" (Rohlf and Marcus 1993). The revolution has matured in recent years (Adams et al. 2004, 2013) and has been accompanied by powerful data acquisition tools such as micro-focal X-ray computed tomography (micro-CT) scanners that allow high-resolution scanning of very small objects, such as small mammal skulls. By linking scans with software programs (such as MATLAB and AVISO), it is now possible to conduct "virtual dissections" of internal structures of the cranium, such as the cochlea, endocranium, nasal capsule, and semicircular canals, and to perform sophisticated analyses of the 3D shape of these minute structures. Such approaches, combined with molecular techniques to sequence functional genes related to different senses, have led to a deeper understanding of the evolution and ecology of sensory systems. Many of these advances have involved small mammals. For a review of some key resources on morphometrics, see Appendix A at the end of this chapter (see also Rohlf and Bookstein 1990, Bookstein 1991, Marcus et al. 1993, 1996, Zelditch et al. 2012, Adams and Otárola-Castillo 2013, Claude 2013).

The last three decades has also seen the rapid development of molecular systematics and the widespread adoption of cladistic methods (defining evolutionary relationships based on shared ancestry) over phenetic methods (defining evolutionary relationships based on overall similarity). As elaborated in this chapter, superimposing small mammal morphological characters onto robust phylogenies can reveal underlying evolutionary adaptive patterns and processes and ancestral character states (Rohlf et al. 1996, Courant et al. 1997, Michaux et al. 2007, Renaud et al. 2018). Furthermore, in small mammal studies where taxonomic instability often prevails, examination of morphological (including craniometric and craniodental) characters remains crucial for field-based and museum-based identification and delimitation of cryptic species (for examples of African rodents, see Monadjem et al. 2015).

Aims

A chapter on morphometrics may seem to be out of place in a book on field ecological methods. However, as we elaborate in the examples given below, morphological data underpinned by robust phylogenies and ecological data provide an effective interpretive framework to investigate ecological phenomena such as evolutionary convergence and phenotypic adaptation to environmental gradients or climate change. This chapter is intended to be a concise primer educating small mammal ecologists and researchers on the use of modern morphological and morphometric tools to support and enrich ecological studies, especially by accurately delimiting species and by illuminating adaptive trends. A basic understanding of the central concepts in morphometry (e.g., homology and allometry) is essential for good practice, and therefore briefly articulated in this chapter, along with a few examples of ecological applications that may be useful to the reader.

Homology

Two structures are called homologous if they represent corresponding parts of organisms that are built according to the same body plan due to shared common ancestry (Van Valen 1982, Wagner 1989). The opposite of homology is homoplasy, which occurs where similarity between structures in different taxa may be due *not* to common descent but rather to evolutionary convergence (where structures come to resemble each other in unrelated taxa due to common evolutionary pressures). In morphometrics, while two structures in different taxa may be homologous, landmarks (defined as standardized and repeatable anatomical points) internal and external to those structures may not be homologous in a strict sense. This has led to a more operational interpretation of homology in the field of morphometrics, compared to the strict evolutionary or "taxic" definition. The distinction between these divergent schools of thought concerning homology, operational and taxic, is discussed in detail by Smith (1990).

Allometry

Allometry is the study of the relationship of body size to shape, anatomy, physiology, and behavior. Julian Huxley and Georges Tessier first coined the term in 1936 in the context of relative growth in the large claw of the male fiddler crab, which grows disproportionately faster than growth in body mass. A full treatment of this massive field is outside the scope of this chapter, but as the basis for any organismal study of small (versus large) mammals, it is useful to have some brief insight into the allometric trajectories of different anatomical and physiological aspects of mammals in general (see below). Also, since size-related shape changes are common in animals, caution should always be taken to consider allometric effects in any study involving species of different sizes. A useful introduction to the topic of allometry is provided by Shingleton (2010) and Gayon (2000) (Appendix A).

Since its original formulation, different types of allometry have been defined (1) between growth stages (ontogenetic allometry), (2) between individuals of one growth stage (static allometry), and (3) between different species (evolutionary allometry). Allometric relationships are typically represented mathematically by the power function $Y = aM^b$ or its linear function when values are log-transformed, $\log(Y) = \text{Log}(a) + b \times \log(M)$, where M is body mass, Y is the variable to be tested, a is a proportionality constant, and b is the scaling coefficient. Where $b = 0$, there is no relationship between a trait and body mass. Where $b = 1$, when traits increase in direct proportion to body mass, this is termed "isometry." Where $b > 1$, when traits increase faster than body mass, this is termed "positive allometry." When traits increase slower than body mass ($b > 0$ and < 1), this is termed "negative allometry." When traits decrease with body mass, i.e., with a negative slope, this is termed "inverse allometry" ($b < 0$). In mammals, heart size scales isometrically with body mass while brain size scales negatively ($b = ¾$, or 0.75). Bone thickness is one trait than scales positively allometrically with body mass ($b = \frac{4}{3}$, or 1.33) in mammals (e.g., compared to small mammals, like rodents and shrews, the bone thickness of large mammals, like elephants, is proportionately greater than explained just by body-size differences). In contrast, whole-animal metabolic rate increases with body size but not as fast as body mass increases, i.e., it is negatively allometric (b = ¾, or 0.75), while mass-specific metabolic rate scales inversely with body mass in mammals (b = −¼, or −0.25), meaning that small mammals like shrews and rodents have a much higher relative (per gram body mass) metabolic rate compared to elephants and other large mammals. Longevity in mammals scales negatively with body mass (b = ¼, or 0.25). Interestingly, the period between two heartbeats in mammals scales to exactly the same coefficient (b = 0.25). As an approximation, in the case of all mammals, from a 25 g rodent that lives 3.5 years to an elephant of 3000 kg that lives 70 years, one lifespan can be equated to about 1.5 billion heartbeats (Shingleton 2010).

External, Craniodental, and Postcranial Anatomy of Small Mammals

Small mammals share a common body form, making it possible to adopt a standard set of external measurements, although there is considerable variation in body size, coloration, and relative size of the tail, limbs, ears, and eyes. For example, in golden moles (order Afrosoricida, family Chrysochloridae), tail, external pinna, and functional eyes are all absent. As described in detail in Chapter 6, weights and standard external measurements can be taken from live, anesthetized, or sacrificed individuals.

Since museum specimens typically retain the entire cranium (mandible or lower jaw and maxilla or upper jaw) including the teeth, and since these are highly complex and variable structures, it is not surprising that craniodental characters and measurements are widely used in the study of small mammal taxonomy, growth, and ecology. Since the teeth and some cranial features (such as the shape of the zygomatic arch) are strongly linked to diet, they can be used to make inferences about diet, bite strength, and so forth. An excellent general guide to the skull, teeth, and postcranial skeleton of mammals can be found in the third edition of *A Manual of Mammalogy* (Martin et al. 2011).

Types of Data

Although external and cranial characters and measurements can be taken for a variety of purposes, such as further ecological or taxonomic study, one of the most common reasons they are taken is to help with the identification of a specimen to species level. Some of the different kinds of nonquantitative (qualitative) and quantitative characters used for species identification and additional analyses are defined below.

Qualitative Characters

Qualitative characters or traits are nonquantitative. While quantitative data can be measured, counted, and expressed with numbers, qualitative data are descriptive and can be categorized based on characteristics. For example, the character of fur color in a group of small mammals could be categorized as black or gray. In small mammals, dental characters, such as the presence or absence of a particular tooth, or a cusp on a tooth, are commonly used. Characters that are unique to a species or group of species are described as diagnostic. As described below, under "Applications of Morphology to Ecological and Taxonomic Studies," such diagnostic traits are useful for designing taxonomic keys as well as for constructing phylogenies using cladistic analysis.

Linear Measurements (Quantitative Data)

A range of standard linear measurements has been used to describe the morphology of small mammals. Standard external measurements such as head and body, tail, ear, and hindfoot lengths are described in Chapter 6. Such measurements are usually taken with a ruler or a pair of calipers. Craniodental and mandibular measurements of small mammals are usually taken with digital calipers having a resolution of at least 0.1 mm, preferably 0.01 mm. To ensure repeatable and accurate measurements, it is worth paying a bit extra for a reliable pair of calipers. Mitutoyo and Tesa both make excellent calipers. Some of the standard cranial measurements taken on small mammals include condylobasal length, basal length, basilar length, greatest skull length, zygomatic width, molar toothrow length, palatal length, post-palatal length, interorbital width, mastoid width, braincase width, and mandible and incisor length (Fig. 14.1, Table 14.1), but researchers often take a wider variety of measurements. In the case of shrews, some measurements will differ. For example, shrews lack zygomatic processes, so the widest skull measurement is bimaxillary width measured across the maxillary processes. Toothrow length in shrews is measured from the anterior surface of the incisor to the posterior surface of the last molar, whereas in rodents, it is more common to report maxillary toothrow length, i.e., excluding the incisors and diastema. While condylobasal length (anterior point of the premaxilla) is commonly recorded in rodents, condylo-incisive length (anterior point of the incisors) is used for shrews.

Fig. 14.1. Dorsal (*A*), ventral (*B*), and lateral (*C*) views of the cranium and lateral view of the mandible (*D*) of a Nile rat (*Arvicanthis niloticus*) showing anatomical landmarks and typical craniodental measurements. Abbreviations of measurements are defined in Table 14.1. *(From Abdel-Rahman 2005)*

A.

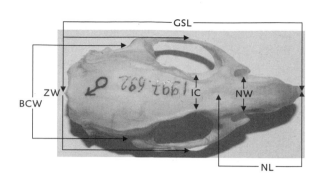

B.

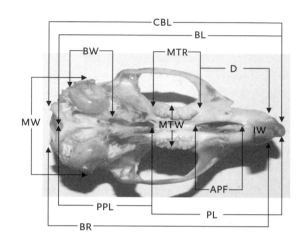

C.

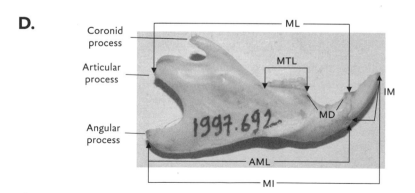

D.

Table 14.1. Craniodental and mandibular measurements illustrated in Figure 14.1 (from Abdel-Rahman 2005)

Maxillary measurements

APF, anterior palatine foramen	Distance from anterior-most to posterior-most part of the palatine foramen
BCW, braincase width	Braincase width measured at dorsal roots of zygomatic arches
BL, basal length	Distance from an anterior edge of the premaxillae to posterior-most projection of occipital condyles
BH, bulla height	Height of bulla measured laterally from dorsal to ventral surface
BR, basilar length	Distance from the posterior surface of incisor at alveolus to anterior-most part of the foramen magnum ventrally
BW, bulla width	Greatest diameter of auditory bulla
CBL, condylobasal length	Distance from an anterior edge of the premaxillae to anterior-most part of the foramen magnum ventrally
D, maxillary diastema length	Distance from cingulum of upper incisor to cingulum of first upper molar (M1)
GSL, greatest skull length	The greatest distance between the posterior-most point of the skull and the anterior-most point of the nasal bone
HBC, depth of the braincase	Greatest depth of the braincase from the top of the parietal bone to the basisphenoid region
IC, interorbital construction	Least distance dorsally between the orbits at right angles to the longitudinal axis of a skull
IL, upper incisor length	Distance from alveolus of upper incisor to tip of incisor
IW, upper incisor width	Width of both upper incisors combined
MTR, maxillary toothrow	Distance from anterior edge of alveolus of first upper maxillary cheek-tooth to posterior edge of last maxillary cheek-tooth
MTW, width across molar toothrows	Greatest width across upper molar toothrows
MW, mastoid width	Greatest width of the skull between the tympanic bullae
NL, nasal length	Greatest length from the anterior to posterior edge of the nasal bones
NW, nasal width	Greatest width of the nasal bones
PL, palatal length	Distance from anterior edge of the premaxillae to anterior-most point of posterior edge of palate
PPL, post-palatal length	Distance from anterior-most point on posterior edge of palate to anterior-most part of the foramen magnum ventrally
ZW, zygomatic width	Greatest distance between the outer margins of the zygomatic arches

Mandibular measurements

AML, ventral mandible length	Distance from posterior-most point of the angular process to ventral surface of alveolus of the lower incisor
IM, lower incisor length	Distance from alveolus of upper incisor to tip of incisor
MD, mandibular diastema length	Distance from alveolus of lower incisor to alveolus of lower first molar
MI, mandible and incisors	Distance from posterior surface of angular bone to tip of incisors
ML, dorsal mandible length	Distance from posterior-most point of the articular process to dorsal surface of alveolus of the lower incisor
MTL, lower toothrow length	Distance from anterior edge of alveolus of first lower maxillary cheek-tooth to posterior edge of last

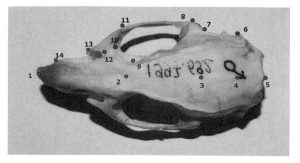 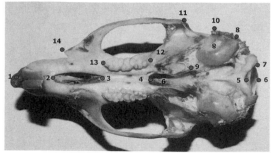

Fig. 14.2. Dorsal (*A*) and ventral (*B*) views of the skull of Nile rat (*Arvicanthis niloticus*) illustrating position of dorsal and ventral landmarks used for a 2D geometric morphometric study of the species *(From Abdel-Rahman 2005)*

Homologous Landmarks (2D and 3D), Semi-landmarks, Outlines, and Surfaces

Linear cranial measurements ("distances") are often highly correlated with each other and strongly influenced by the general size of the cranium. Multivariate analyses of distance measurements cannot effectively partition shape and size. Landmarks have the advantage of being "size-free" variables that overcome some of the problems of linear measurements. Landmarks in morphometrics are sets of Cartesian coordinates in two (x, y) or three (x, y, z) dimensions placed on an object or image of an object and used to capture the shape of a particular structure or set of structures. Such shape information can be used for further ecological, taxonomic, or phylogenetic studies, as explained below. Figure 14.2 shows a typical set of landmarks placed on a dorsal and ventral image of a rodent skull. There are three recognized categories of landmarks (Bookstein 1991). Type 1 landmarks define intersections, e.g., between three sutures. Type 2 landmarks include specific points, such as tips of structures or local minima and maxima of curvature. Type 3 landmarks are defined in terms of a point "furthest away" from another point. Type 1 landmarks are better (more repeatable) than type 2, which are better than type 3. In addition to landmarks, there are "semi-landmarks," points whose position along a curve is fixed but that provide information about curvature in two or three dimensions. Semi-landmarks can be useful for curved surfaces where discrete and homologous landmarks are not possible, e.g., the cochlea (Fig. 14.3), the bulla, etc. Outlines and surfaces are less commonly used in geometric morphometrics than are landmarks. In the case of round or curved structures without definite points or intersections, such as rodent mandibles and the occlusal surfaces of individual teeth, outlines may be more appropriate than landmarks.

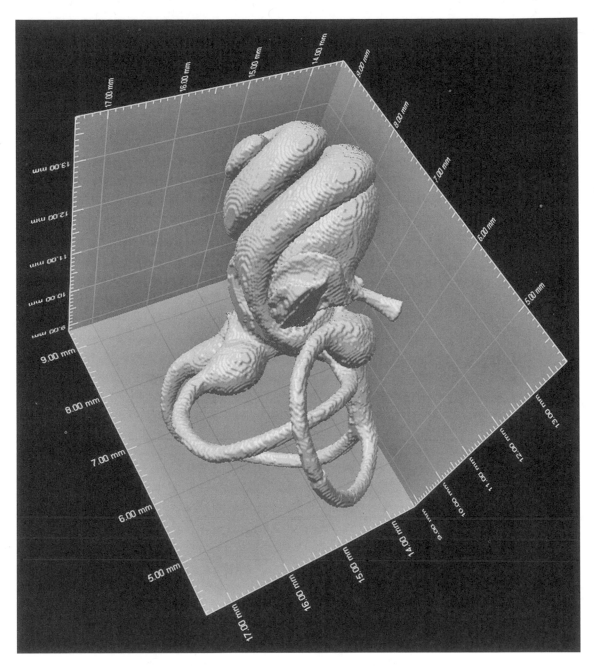

Fig. 14.3. Example of 3D image of cochlea and semicircular canals from an African vlei rat (*Otomys auratus*) reconstructed using micro-CT imaging and the AVISO software at the South African Nuclear Energy Corporation (NECSA) facility in Pelindaba, South Africa, from a museum specimen from the Ditsong National Museum of Natural History, Pretoria, South Africa (Catalog No. 19287) *(Photo: Aluwani Nengovhela from Nengovhela 2018)*

Accounting for Sources of Variation

When conducting a morphometric study, the purpose of the investigation should be clearly stated, and any sources of error that may compound the main effect under investigation should be properly accounted for to prevent biases affecting conclusions. In all morphometric studies, measurement error (due to accuracy and precision) needs to be considered. Population studies or taxonomic studies should account for four common sources of intrapopulation variation: individual variation, ontogenetic (age) variation, seasonal variation, and secondary sexual dimorphism. Since geographic variation within a species may mask or bias differences between species, taxonomic studies should additionally consider interpopulation (geographic) variation. To statistically account for different sources of variation, sufficient and unbiased representation of the different groups to be tested (whether age classes, sexes, or geographical localities) in the final study sample is critical. For example, a study of geographical variation in a species that only samples a very small part of the species' range is clearly inadequate. A study of sexual dimorphism where the age distribution of the males and females is significantly skewed would also be flawed. Likewise, a study attempting to morphometrically diagnose two geographically highly variable species by considering just one locality of each is flawed. Such information could hardly be used to develop a key to identify specimens obtained from throughout the ranges of both species. Sample sizes between groups should be as balanced as possible, although this is often a constraint when using museum collections, which are notoriously biased, often reflecting the favorite or most accessible collecting sites used by a particular collector (Maddock and Du Plessis 1999). The analysis of different sources of morphometric variation needs to be logical and progressive, e.g., measurement and individual error needs to be assessed first, before consideration of ontogenetic and sexual variation (intrapopulation), which should be addressed before consideration of geographic (interpopulation) variation, which should be addressed before consideration of taxonomic variation (e.g., species differences).

Accuracy measures the extent to which measurements represent a quantity's true value. Precision represents the similarity between repeated sets of measurements (repeatability). Since all scientific studies should be repeatable by other researchers, both precision and accuracy are important to any studies of morphometric variation in small mammals. A consistent measurement error (e.g., parallax bias in landmarks obtained from digital images placed too close to the object: Mullin and Taylor 2002) may result in low accuracy but high (or low) precision. The coefficient of variance (CV), which is the standard deviation expressed as a percentage of the mean, is often used as an index of the usefulness of a morphometric variable. However, this index may be a function of high intrapopulation variability or low precision or accuracy of the measurement system. Measurement error can be isolated by taking multiple measurements of the same set of individuals at different times by the same recorder (Lessells and Boag 1987). Intraindividual variation can be equated with repeatability or precision (Wolak et al. 2012). In geometric morphometric studies, this can be due either to biased capturing of images by the imaging device or to digitizing errors in placing landmarks on the same image. Measurement error (%) can be calculated as

$$\% \, ME = \left[\frac{s^2_{within}}{s^2_{within} + s^2_{among}} \right] \times 100,$$

where s^2_{within} represents the within-mean square effect (or the within-specimen variance) and s^2_{among} represents the among-mean square effect (or the among-specimen variance). Using such an approach, Abdel-Rahman et al. (2009) found that measurement error (differences within individuals) in collecting landmarks using a digital camera (both from landmark placement on the images and retaking images from the same individual) was <8% of the total sample variance. Values of <15% are considered reliable for estimating biological trends (Bailey and Byrnes 1990).

In most morphometric studies of small mammals, it is not possible to obtain absolute ages of individuals, and it is necessary to estimate the relative age of individuals, usually with the aim of removing juvenile and subadult individuals that would bias results. A number of methods have been used to "age" small mammals, including body weight, skull size, eye-lens weight, tooth eruption, and tooth wear (see Morris 1972, Taylor et al. 1985, Chapter 6). Most commonly in morphometric studies of small mammals, skulls can be sorted into relative age classes based on detailed tooth wear and eruption (Fig. 14.4). Significantly variant tooth-wear/age classes should be removed prior to subsequent analysis, so as not to bias the results of subsequent analyses, e.g., to determine species differences.

Sexual dimorphism is often present in small mammals and can be tested using univariate (e.g., analysis of variance, or ANOVA) and multivariate techniques (e.g., multivariate analysis of variance, or MANOVA, and canonical variates analysis). Where significant differences occur between the sexes, subsequent analyses should be performed separately on separate samples for males and females. Since growth rates may also vary between sexes, a two-way ANOVA can be used to simultaneously test for variation due to age and sex as well as the interaction between them. Such an approach was used to compare growth rates of male and female Samango monkeys (*Cercopithecus*

Fig. 14.4. Right maxillary toothrow of a multimammate mouse (*Mastomys coucha*) illustrating seven tooth-wear classes of increasing wear (relative age) arranged from Class I to Class VII *(Photos: Seth Eiseb from Eiseb 2015)*

albogularis) in southern Africa (Dalton et al. 2015) and can easily be applied to small mammals as well. In some cases, body weight or other measurements may vary seasonally due to annual reproductive change or population structure, as found in pouched mice in South Africa, *Saccostomus campestris* (Ellison et al. 1993). Any future analysis of body size in this species should then group samples into seasonal groups, or only consider samples from a single season.

Multivariate Morphometrics

During the 1960s and 1970s, biometricians developed a wide range of multivariate methods to analyze morphometric variables, usually linear distances between standard landmarks. This approach has come to be known as traditional morphometrics (Marcus 1990) or multivariate morphometrics (Blackith and Reyment 1971). Both univariate and multivariate techniques can be employed. One of the most commonly used univariate techniques in morphometric studies is ANOVA, to test for group differences (e.g., sexes, age groups, populations, or species). Two-way ANOVA is often used where two group effects and their interactions are considered simultaneously (e.g., sex and age).

Multivariate methods comprise a set of tools for analyzing multiple variables in an integrated and powerful way (see also Chapter 10). They are frequently employed in morphometric studies and better elucidate differences compared with univariate methods that analyze dependent variables one at a time. Common multivariate techniques in morphometric studies include exploratory methods, such a principal component analysis (PCA) and cluster analysis (CA), and statistical methods, such as discriminant functions analysis (DFA). Table 14.2 provides a general guide as to the choice of multivariate

Table 14.2. Guide as to the choice of multivariate methods depending on the research focus and the nature of the independent and dependent variables

Research focus	IVs (No. & scale)	DVs (No. & scale)	Multivariate method
Group differences			
	1+ categorical & continuous	1 continuous	ANCOVA
	1+ categorical	2+ continuous	MANOVA
	2+ continuous	1+ categorical	DFA
	1+ categorical or continuous	1 categorical	LR
Correlational			
	2+ continuous	1 continuous	MR
	2+ continuous	2+ continuous	CC
	—	2+ continuous	PCA & FA

Note: Scale and number of independent (IV) and dependent (DV) categorical or continuous variables. + indicates 1 or more; ANCOVA = analysis of covariance; MANOVA = multivariate analysis of variance; DFA = discriminant function analysis; LR = logistic regression; MR = multiple regression; CC = canonical correlation; PCA/FA = principal components/factor analysis

method in a particular study, depending on the research question and the nature of the independent and dependent variables studied.

PCA is a commonly used multivariate method that summarizes information from a number of morphometric variables in reduced dimensionality, making it easier to detect patterns among individuals or groups. The technique is summarized very succinctly and clearly by Hammer (2002) as follows:

> Principal components analysis (PCA) is a method that produces hypothetical variables or components, accounting for as much of the variation in the data as possible (Jolliffe 1986). The components are linear combinations of the original variables. This is a method of data reduction that makes it possible to present the most important aspects of a multivariate dataset in two dimensions, in a coordinate system with axes that correspond to the two most important (principal) components. In other words, PCA represents a way of projecting points from the original, high-dimensional variable space onto a two-dimensional plane, with minimal loss of information. In addition, these principal components may be interpreted as reflecting underlying morphological variables with a biological meaning.

Statistical programs also present the "loadings" of components, indicating how much each original variable contributes to the components. In morphometric studies, the first component, PC1, is typically interpreted as a "size vector" since the loadings of all of the morphometric variables on PC1 are typically of similar magnitude and sign (positive), indicating that individuals having high PC1 scores (larger sized) also have high values of all the original variables.

A simple and free statistical package that does both univariate and multivariate statistics is PAST (Hammer et al. 2001) (Appendix A). The package also does useful data manipulation, transformation, and plotting functions, as well as some geometric morphometric functions using landmarks, performing Procrustes superimposition, and so on. An excellent and simple 50-page document, "Morphometrics—Brief Notes," is provided by Øyvind Hammer (Hammer 2002). This is strongly recommended reading for anyone wishing to do morphometric work. This resource covers everything from data acquisition, statistical assumptions, univariate and multivariate methods, traditional landmark-based approaches, allometry, and statistical modeling. For those comfortable with the R statistical platform, there are several packages available, such as the "vegan" package, as well as online guides and books (Appendix A).

The Geometric Morphometric Revolution

Geometric morphometrics (GM) represents a radical synthesis between multivariate statistical methods and geometry in the form of transformation grids popularized by Thompson's (1917) *On Growth and Form*. The approach allows the geometry of shape changes to be analyzed and visualized with respect to Cartesian coordinates that are invariant to scale, orientation, and location. At its essence, the power in GM is its ability to fully separate size and shape, and to further decompose shape into two distinct entities, uniform (affine) and nonuniform (nonaffine) shape. The nature of uniform and nonuniform shape can be understood with reference to the transformation grids of Thompson (1917), shown in Figure 14.5. Thus, the fish genus *Argyropelecus* represents the standard (average) or "consensus" shape represented by a grid having exactly square (i.e., equal sided) cells. Two kinds of uniform deformation can be recognized where changes are constant between grid cells, "shear" and "dilation." These effects are illustrated by the fish genera *Sternoptyx* and *Diodon*, respectively, in Figure 14.5. Nonuniform changes occur where deformations are localized for particular regions, as illustrated by the fish genus *Orthagoriscus* in Figure 14.5, where the anal fins and dorsal fins are greatly expanded dorsoventrally relative to the anterior regions of the fish. These shape changes illustrated manually by Thompson can now be computed mathematically using the thin plate spline (Fig. 14.5).

Acquisition of either 2D or 3D landmarks necessary for GM can be undertaken by a range of different imaging devices, including scanning devices, digital cameras (see Fig. 14.2) and more recently, advanced micro-CT scan-

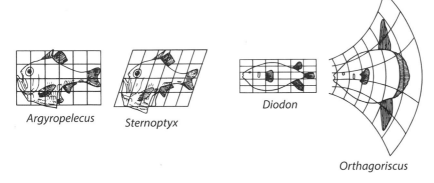

Fig. 14.5. Examples of deformation grids from D'Arcy Thompson (1917) used to illustrate uniform shape changes between one reference (*left*) and three target (*right*) genera of fish. *(Redrawn from MacLeod, https://www.palass.org/publications /newsletter/palaeomath-101/palaeomath-part-19-shape-models-ii-thin-plate-spline)*

ning in 3D. Using micro-CT scanning, the structure of interest (e.g., skull or an external or internal part thereof such as the cochlea, braincase, or nasal capsule) can be "virtually dissected" and saved as a 3D mesh file (e.g., ply file) (see Fig. 14.3 for such a reconstructed image of the cochlea and semicircular canals of a rodent of the genus *Otomys*). Micro-CT scanning is a nondestructive imaging technique that uses X-ray energy to create volumetric reconstructions of a specimen. Typically, around 1000 or more X-ray slices are obtained horizontally for a small rodent skull. These are used to build a composite image that can be manipulated with expensive proprietary software such as AVISO and VGSTUDIO MAX. Many facilities around the world own micro-CT scanners, and some provide a service to conduct micro-CT scanning and image manipulation to researchers at a nominal cost. Images obtained from such facilities can be exported to formats that can be analyzed later. Once 3D coordinates are obtained from 3D images, they can be analyzed in exactly the same way as 2D coordinate data (see below).

Shape and size differences can be partitioned in a particular structure among individuals by superimposing landmark configurations derived from their images (in 2D or 3D) (Fig. 14.2), and depicting, both statistically and visually, individual deformations and trends relative to a consensus or reference configuration by means of the so-called thin plate spline method (Marcus et al. 1996, Adams et al. 2004, 2013, Claude 2013). A number of different methods are used to translate, rotate, and scale configurations of landmarks in 2D or 3D representing a set of individuals. This process effectively partitions size and shape. One of the most commonly used methods, the generalized Procrustes analysis (GPA) method, uses least squares to compute a consensus

mandibles of Old World moles (Talpidae) (Rohlf et al. 1996), as well as African root rats (Spalacidae) (Corti et al. 1996) and various genera of Muridae (Fadda and Corti 1998, 2000, 2001, Rohlf and Corti 2000, Corti et al. 2004). These studies all emphasized the superior approach of GM compared with traditional methods in distinguishing shape from size when defining species limits or investigating ecological correlates of skull size and shape in these taxa. Claude (2013) tested the power of different GM methods—log-shape ratios on body measurements, elliptic Fourier analyses on teeth outlines, and Procrustes superimposition on skull coordinates—in discriminating between two commensal rat species, *Rattus exulans* and *R. tanezumi* in Thailand.

Functional Morphology and Adaptive Trends

As with most vertebrates, mass (body size) of small mammals, and different proxies of this (such as body length or skull length), have been found to be functionally important and strongly correlated with different ecological variables. For example, in widespread species and genera, larger-sized individuals of a species, or members of related species, occur at colder climates associated with higher latitudes and altitudes. This trend is typically explained by Bergmann's rule (Bergmann 1847), which postulates that smaller individuals with a higher surface area relative to volume lose heat more efficiently in warm climates (at low latitudes and altitudes), whereas the opposite is true in colder environments (at high latitudes and altitudes), where a lower surface-area-to-volume ratio (larger body size) allows better heat conservation. Bergmann's rule has also been invoked to explain temporal decreases in body size due to climate warming, or the so-called "third universal response to global warming" (Gardner et al. 2011). Data on cranial length from museum specimens collected over 100 years confirmed this pattern in at least two southern African laminate-toothed rat (*Otomys*) species (Nengovhela et al. 2015). However, in a global review of temporal trends in body or cranial size in 50 rodent species (Nengovhela et al. 2020), only 13 species showed a decline in size over time, leading the researchers to conclude that the "third universal response" rule may not be so universal in rodents, but that largely unexplained attributes result in different body-size responses (increases, decreases, and no changes) to climate warming in rodents.

Apart from body and cranial size, different cranial measurements and ratios can be used as proxies for functional parameters in small mammals. For example, auditory bullar hypertrophy (increase in size) in rodents is associated with convergent adaptations to a desert environment in gerbils (Muridae: Gerbillinae), kangaroo rats (Heteromyidae), and laminate-toothed rats (Muridae: Otomyini) (Nengovhela et al. 2019). This association may be due to enhanced auditory sensitivity in desert rodents for either predator avoidance or enhanced conspecific communication (see also Dempster 2018). Mandibular morphology has been shown to be related to bite force in shrews (Soricidae), which in turn is correlated with aridity (Carraway and Verts 1994).

Due to convergent evolution, caution should be used when employing morphological evidence for phylogenetic inference. However, when shape information is superimposed on a phylogenetic framework, this is a powerful tool for distinguishing convergent from divergent traits. For example, Rohlf et al. (1996) found that nonuniform (nonaffine), but not uniform (affine), shape differences in the cranium correctly identified New and Old World phylogenetic clades of moles (Talpidae). Courant et al. (1997) found that especially the lateral profile of the crania of voles and lemmings was strongly correlated with soil substrate hardness and the ability to burrow (burrowing species had more angular lateral profiles). By superimposing shape information on an independently derived phylogeny, they could identify four "ecological events" leading to morphological convergence associated with different substrates and burrowing behavior (Fig. 14.6).

Outlines and surfaces are less commonly used in GM than are landmarks. In the case of round or curved structures without definite points or intersections, such as rodent mandibles and the occlusal surfaces of individual teeth, outlines may be more appropriate than landmarks. By comparing function(s) of a curve, and not the points collected on the outline, mathematical Fourier descriptor methods can allow investigation of shapes with few homologous landmarks (Rohlf and Archie 1984). Both radial and elliptical Fourier transformations can be performed. For detailed explanations of these techniques and their comparison with landmark-based methods, see Sheet et al. (2006) and Renaud et al. (2018).

A good example of an outline analysis relating small mammal form to ecological function is the study of Michaux et al. (2007), where mandible outlines of 65 species of 45 genera of murine rodents were analyzed by elliptical Fourier analysis, revealing effects of both phylogeny and ecology on mandible shape. PCA of elliptical Fourier coefficients derived from the analysis allowed variation in mandible shape between species to be summarized in two dimensions (PC1 and PC2). From a given set of Fourier coefficients, an outline of the mandible can be reconstructed using an inverse method (Rohlf and Archie 1984), and this allows a visualization of the mandible-shape changes involved along each principal component. By visualizing these reconstructed changes for PC1 and PC2 in relation to the scatter of data, a relationship could be observed between the diet and mandible shape of different groups. For example, negative scores on PC1 reflected massive mandibles, associated with

APPENDIX A. USEFUL INTERNET LINKS FOR MORPHOMETRIC TECHNIQUES

Topic	Originator/ reference	URL
Online journal issue on morphometrics	*Hystrix* journal	http://www.italian-journal-of-mammalogy.it/Issue-1-2013,2854
"Geomorph" R package and general instruction about GM	Emma Sharratt, University of Adelaide	http://www.emmasherratt.com/news—musings
Allometry	Shingleton (2010)	https://www.nature.com/scitable/knowledge/library/allometry-the-study-of-biological-scaling-13228439
"PAST" statistical software	Hammer et al. (2001)	https://past.en.lo4d.com/windows
Principal component analysis (using R)	Alboukadel Kassambara PhD, Bioinformatics R&D Scientist, HalioDx, Marseille	http://www.sthda.com/english/articles/31-principal-component-methods-in-r-practical-guide/118-principal-component-analysis-in-r-prcomp-vs-princomp/
Thin plate spline method of GM	Norman Macleod, Natural History Museum, London	https://www.palass.org/publications/newsletter/palaeomath-101/palaeomath-part-19-shape-models-ii-thin-plate-spline

APPENDIX B. KEY TO FAMILIES OF AFRICAN RODENTS (MODIFIED FROM MONADJEM ET AL. 2015)

1 Undersurface of tail with obvious large scales at its based; flight membrane present (except *Zenkerella*); strictly arboreal . . . . Anomaluridae

 Undersurface of tail without obvious scales at its base 2

2 Body covered in spines; typically large (mass > 2 kg; HB > 500 mm) Hystricidae

 Body not covered in spines; small to large. 3

3 Bipedal with enormous hindlimbs in proportion to forelimbs (at least five times as long) 4

 Not bipedal with hindlimbs similar in length to forelimbs. 5

4 Large (mass > 1 kg; HB > 300 mm). . . . . Pedetidae

 Small (mass < 100 g; HB < 200 mm). . . . . Dipodidae

5 Body large (mass > 1.5 kg; HB > 400 mm) and stocky with short tail (30%–75% of HB) 6

 Body not as above 7

6 Body beaver-shaped with long, shaggy fur and longer tail (ca. 70%–75% of HB); upper incisors ungrooved; feral populations in Kenya only . . . . . . Myocastoridae

 Body with shorter fur and short tail (30%–40% of HB); upper incisors with deep longitudinal grooves . . . . .Thryonomyidae

7 Fossorial with mole-like appearance; short tail (<50% of HB), small eyes, short ears 8

 Not fossorial. 9

8 Hard palate extends well beyond posterior margin of last upper molar . . . .
 . . . . Bathyergidae

 Hard palate does not extend beyond posterior margin of last upper molar . . . . . .
 . . . . Spalacidae

9 Guinea pig shaped with stocky body, soft fur, short tail (30%–45% of HB); strictly rupicolous. . . . . Ctenodactylidae

 Not as above. 10

10 Three upper cheekteeth; typically rat-like with long, sparsely furred tail 13

 Four upper cheekteeth; typically squirrel- or dormouse-like in appearance . . . 11

11 Tail sparsely covered in long hairs; strictly rupicolous; restricted to arid Namibia (marginally into South Africa and Angola). .Petromuridae

 Tail bushy 12

12 Smaller (mass < 100 g; HB < 160 mm); dormouse-like, typically with gray pelage; uniform dark-colored tail (with or without white tip); skull with interorbital constriction; postorbital process absent; nocturnal . . . . Gliridae

 Larger (mass > 40 g; HB > 100 mm); squirrel-like with pelage not uniform gray; tail typically grizzled or banded; skull without interorbital constriction; postorbital process well developed; diurnal . . . . Sciuridae

13 First lobe of M^1 with three cusps (t1 present)* . . . . Muridae

 First lobe of M^1 with two cusps (t1 absent). .Nesomyidae

*Except for the Gerbillinae, *Deomys* and *Leimacomys* (there are in fact no unique morphological features that distinguish these two families, although they are clearly separable on molecular grounds).

Michaux, J., P. Chevret, and S. Renaud. 2007. "Morphological diversity of Old World rats and mice (Rodentia, Muridae) mandible in relation with phylogeny and adaptation." *Journal of Zoological Systematics and Evolutionary Research* 45: 263–279. https://doi.org/10.1111/j.1439-0469.2006.00390.x.

Monadjem, A., P. J. Taylor, C. Denys, and F. P. D. Cotterill. 2015. *Rodents of Sub-Saharan Africa: A Biogeographic and Taxonomic Synthesis*. Berlin: De Gruyter.

Morris, P. 1972. "A review of mammalian age determination methods." *Mammal Review* 23: 69–104.

Mullin, S. K., and P. J. Taylor. 2002. "The effects of parallax on geometric morphometric data." *Computers in Biology and Medicine* 32 (6): 455–464.

Nengovhela, A. 2018. "3D cranial morphometry, sensory ecology and climate change in African rodents." PhD thesis, dual degree from Paul Sabatier University of Toulouse, France, and the University of Venda, South Africa.

Nengovhela, A., R. M. Baxter, and P. J. Taylor. 2015. "Temporal changes in cranial size in South African vlei rats (*Otomys*): Evidence for the 'third universal response to warming.'" *African Zoology* 50 (3): 233–239. https://doi.org/10.1080/15627020.2015.1052014.

Nengovhela, A., J. Braga, C. Denys, F. De Beer, C. Tenailleau, and P. J. Taylor. 2019. "Associated tympanic bullar and cochlear hypertrophy define adaptations to true deserts in African gerbils and laminate-toothed rats (Muridae: Gerbillinae and Murinae)." *Journal of Anatomy* 234: 179–192. https://doi.org/10.1111/joa.12906.

Nengovhela, A., C. Denys, and P. J. Taylor. 2020. "Life history and habitat do not mediate temporal changes in body size due to climate warming in rodents." *PeerJ* 8: e9792. http://doi.org/10.7717/peerj.9792.

Renaud, S., R. Ledevin, B. Pisanu, J. Chapuis, P. Quillfeldt, and E. A. Hardouin. 2018. "Divergent in shape and convergent in function: Adaptive evolution of the mandible in sub-Antarctic mice." *Evolution* 72: 878–892. https://doi.org/10.1111/evo.13467.

Rohlf, F. J., and J. W. Archie. 1984. "A comparison of Fourier methods for the description of wing shape in mosquitoes (Diptera: Culicidae)." *Systematic Zoology* 33: 302–317.

Rohlf, F. J., and F. L. Bookstein, eds. 1990. *Proceedings of the Michigan Morphometrics Workshop*. Ann Arbor: The University of Michigan Museum of Zoology.

Rohlf, F. J., and M. Corti. 2000. "Use of two-block partial least-squares to study covariation in shape." *Systematic Biology* 49 (4): 740–753.

Rohlf, F. J., A. Loy, and M. Corti. 1996. "Morphometric analysis of Old World Talpidae (Mammalia, Insectivora) using partial-warp scores." *Systematic Biology* 45: 344–362.

Rohlf, F. J., and L. F. Marcus. 1993. "A revolution in morphometrics." *Trends in Ecology and Evolution* 8: 129–132.

Sheet, H. D., K. M. Covino, J. M. Panasiewicz, and S. R. Morris. 2006. "Comparison of geometric morphometric outline methods in the discrimination of age-related differences in feather shape." *Frontiers in Zoology* 3: 15.

Shingleton, A. 2010. "Allometry: The study of biological scaling." *Nature Education Knowledge* 3 (10): 2.

Smith, G. R. 1990. "Homology in morphometrics and phylogenetics." In *Proceedings of the Michigan Morphometrics Workshop*, edited by F. J. Rohlf and F. L. Bookstein, 325–338. Ann Arbor, MI: The University of Michigan Museum of Zoology.

Taylor, P. J., J. U. Jarvis, T. M. Crowe, and K. C. Davies. 1985. "Age determination in the Cape molerat *Georychus capensis.*" *South African Journal of Zoology* 20: 261–267.

Taylor, P. J., L. A. Lavrenchenko, M. D. Carleton, E. Verheyen, N. Bennett, and C. Oosthuisen. 2011. "Specific limits and emerging diversity patterns in east African populations of laminate-toothed rats, genus *Otomys* (Muridae: Murinae: Otomyini): Revision of the *Otomys typus* complex." *Zootaxa* 3024: 1–66.

Taylor, P. J., S. Maree, F. P. D. Cotterill, A. D. Missoup, V. Nicolas, and C. Denys. 2014. "Molecular and morphological evidence for a Pleistocene radiation of laminate-toothed rats (*Otomys*: Rodentia) across a volcanic archipelago in equatorial Africa." *Biological Journal of the Linnean Society* 113: 320–344.

Taylor, P. J., S. Maree, J. van Sandwyk, J. C. Kerbis Peterhans, W. T. Stanley, E. Verheyen, et al. 2009. "Speciation mirrors geomorphology and palaeoclimatic history in African laminate-toothed rats (Muridae: Otomyini) of the *Otomys denti* and *O. lacustris* species-complexes in the 'Montane Circle' of East Africa." *Biological Journal of the Linnean Society* 96: 913–941.

Thompson, D. 1917. *On Growth and Form.* Cambridge: Cambridge University Press.

Van Valen, L. 1982. "Homology and causes." *Journal of Morphology* 173: 305–312.

Wagner, G. P. 1989. "The biological homology concept." *Annual Review of Ecology and Systematics* 20: 51–69.

Wolak, M. E., D. J. Fairbairn, and Y. R. Paulsen. 2012. "Guidelines for estimating repeatability." *Methods in Ecology and Evolution* 3 (1): 129–137.

Zelditch, M. L., D. L. Swiderski, and H. D. Sheets. 2012. *Geometric Morphometrics for Biologists: A Primer,* 2nd ed. London: Elsevier, Academic Press.

mammals. Given the rapid progression toward genome-scale approaches, we then give an overview of methods for "genome reduction" that are likely to make major contributions to the study of small mammals. We also provide further details on environmental DNA (eDNA) (see also Chapter 2) for those interested in advancements in these approaches. We include a primer of population genetics, focusing on assumptions for using molecular markers and measuring genetic variation, and we introduce the concept of effective population size. Many of the applications of genetic information are not emphasized herein, but most applications are touched upon in other chapters (e.g., passive detection, Chapter 2; sampling for genetics, Chapter 4; studying movement, Chapter 8; use of genetic markers in estimating demographics, Chapter 9; and studying diet, Chapter 11).

Marker Types

The term "genetic marker" typically refers to a DNA sequence, or locus, typically with a known chromosomal location, although many genetic markers utilized in nonmodel species (i.e., those without a reference genome) are anonymous (i.e., they can be targeted using polymerase chain reaction primers but their chromosomal location is unknown). Different classes of markers convey different types of information, whether at the population or individual level; thus, it is important to have an understanding of the basic properties of commonly used marker types. The advent of polymerase chain reaction (PCR) in the 1980s (Saiki et al. 1985) revolutionized our ability to study DNA markers by enzymatically amplifying short stretches of DNA sequence from targeted regions of the nuclear or mitochondrial genomes from even moderately degraded samples. Below we briefly review the key pros and cons of using select genetic markers to study small mammals. We do not spend time on other historically important techniques, such as restriction fragment length polymorphisms (RFLPs), amplified fragment length polymorphisms (AFLPs), or allozymes, given that their use in such studies on small mammals has become largely obsolete.

Sex-Linked Markers

In ecology it is often important to know the sex of an individual, such as when studying population demographics, sex distributions, or behavioral studies. Markers linked to mammal sex chromosomes allow for identification of sex from samples like feces or hair. This has been of value in passively sampled studies of mammals, for example, where sex cannot be determined visually due to the animal being difficult to observe or capture (Ortega et al. 2004), in

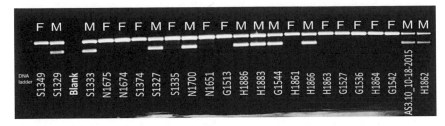

Fig. 15.1. Agarose gel image of sex-specific PCR primers used to distinguish male and female beach mice (*Peromyscus polionotus* spp.). A two gene target approach (Bryja and Konecny 2003) was used to target the *Sry* gene found only on the male Y chromosome, and a second set of primers that co-amplifies related zinc-finger genes located on both Y and X chromosomes. The presence of two bands indicates that *Sry* gene was present, indicating male sex. Individual labels are sample identifiers.

forensic studies requiring genetic determination on juvenile skeletons (Gibbon et al. 2009), or for studies of litter sex ratios and life history variation (Gaukler et al. 2016).

In mammals, PCR of Y-chromosomal DNA markers and autosomal or X-chromosome DNA have been used to determine individual sex (Griffiths and Tiwari 1993, Kohn and Wayne 1997). This can often be done inexpensively by simply examining banding patterns visualized using agarose gel electrophoresis (Fig. 15.1); samples yielding products from both markers are deemed to be male (the heterogametic sex in mammals), and those with only one visible band are female. PCR primers have been designed to amplify a short (~200–250 nucleotides) stretch of the *Sry* gene in a large proportion of mammal species (Griffiths and Tiwari 1993). The *Sry* gene is a widely used Y-linked gene found in most placental (Sinclair et al. 1990) and marsupial (Foster et al. 1992) mammals. Another commonly targeted gene family are gametologous genes, including zinc-finger (ZFY) and amelogenin (AMGY) genes (Shaw et al. 2003).

Although the sequencing of the sex chromosome in small mammals has been underrepresented relative to large mammals, researchers have designed "universal" primers for sex identification (Griffiths and Tiwari 1993). It is important to note that some small mammals have lost their Y chromosomes (Mulugeta et al. 2016), making *Sry* of limited use in some, for example, Ryukyu spiny rat (*Tokudaia osimensis*) (Arakawa et al. 2002, Kuroiwa et al. 2011) and mole voles (*Ellobius* spp.) (Mulugeta et al. 2016). In marsupials, the ZFY and AMGY genes are not located on sex chromosomes, precluding their use as sex markers. For a good place to start with respect to using molecular markers for your species, consult Hrovatin and Kunej (2018), who provide a catalog of molecular sex determination in mammals.

Mitochondrial DNA

Mitochondrial DNA (mtDNA) has been by far the most popular genetic marker to study molecular diversity and evolution in animals over the last three-plus decades (Galtier et al. 2009). Because the mitochondrial genome is maternally inherited in mammals, there is no genetic recombination during meiosis, unlike nuclear DNA (nDNA); thus, mtDNA is readily assayed using direct sequencing. Such uniparental inheritance also results in mtDNA having a small effective population size (see below) relative to the nuclear genome, making it generally more suitable for tracking demographic changes than are nuclear sequences. Also, it is characterized by a relatively high mutation rate, and it has a high copy number in most cell types, making it relatively easy to isolate from small or degraded tissue samples. Furthermore, conserved portions of the mtDNA genome have allowed for the development of "universal" primers that reliably amplify across species, greatly reducing the need for research and development of markers prior to initiating a study on a new organism.

The most common applications of mtDNA are for phylogeographic studies aimed to understand the historical forces that shaped the distribution of genetic lineages within species. For example, mtDNA sequence data has been used to reconstruct the geographic origins and dispersal of many "pest" species, including black rats (*Rattus rattus*) (Aplin et al. 2011) and house mice (*Mus musculus*) (Rajabi-Maham et al. 2008). MtDNA has provided substantial insight into the historical impacts of climate change on small mammal species and communities: for example, that Arctic populations of collared lemmings (*Dicrostonyx torquatus*) underwent repeated regional extinctions and recolonizations during the late Pleistocene (Brace et al. 2012).

Despite these advantages, mtDNA (particularly when examined in lieu of other marker types) has some limitations. Evidence for deviations from strict maternal inheritance in mammals and questions about its strict neutrality (Galtier et al. 2009) have reduced the emphasis of utilizing mtDNA on its own in studies of population structure and demographic history. In addition, due to its haploid inheritance, the mitochondrial genome represents a single genetic locus, which is potentially problematic for accurately reconstructing phylogenetic trees or estimating demographic population histories (Maddison and Knowles 2006). For example, Austin et al. (2018) found no evidence of population differentiation between subspecies of fox squirrels in south Florida based on mtDNA sequences; however, nuclear microsatellites (see below) found very strong evidence for differentiation corresponding to an ecological boundary. The sole reliance on mtDNA in this case would have had profoundly negative management implications for the big cypress fox squirrel

(*Sciurus niger avicennia*). For those considering using mtDNA to study population history or phylogeographic- or phylogenetic-scale questions, we recommend reading Ballard and Whitlock (2004), Ballard and Rand (2005), and Galtier et al. (2009) for overviews of the characteristics and utility of mtDNA.

Microsatellites, or Simple Sequence Repeats

Microsatellites are found throughout the mammalian nuclear genome and are characterized by variation in the number of tandem repeats. Tandem repeats consist of a core DNA motif, typically two to six nucleotides long, that is replicated in succession. For example, a trinucleotide repeat of "GCA" that is repeated six times in a row (e.g., . . . TAGCAGCAGCAGCAGCAG-CACCA . . .) on one chromosomal copy would be characterized as $[GCA]_6$, whereas the second allele present in an individual at this marker might be $[GCA]_{10}$. The number of repeats defines the allele, and the number of alleles at a locus can be as high as 40 (Selkoe and Toonen 2006).

Microsatellites have been the marker of choice since the early 1990s, in part due to their high allelic variation (polymorphism), their relative ease of discovery, and the ability to genotype without directly sequencing each individual (which is relatively costly). Microsatellites typically mutate by addition or loss of repeats, and thus alleles can be identified based on the size of a PCR-amplified product that is determined using capillary electrophoresis. Their high polymorphism is due to their high mutation rates (10^{-4} to 10^{-3} mutations per locus per generation) within and among loci (e.g., Dallas 1992). Because microsatellites have typically been scored via PCR amplicon size (rather than their DNA makeup directly), allele size homoplasy (alleles that are identical in state but are not identical by descent) is likely nontrivial in most species (Estoup et al. 2002) and is a major limitation of this marker type.

This high polymorphism has made microsatellites popular for detecting patterns of individual dispersal and population structure over fine temporal and spatial scales (Austin et al. 2015, 2018), as well as for kin structure (Cutrera et al. 2005) and paternity analysis (e.g., Gryczyńska-Siemiątkowska et al. 2008, Kozakiewicz et al. 2009). Because microsatellites are generally found in short stretches of DNA (tens to hundreds of base pairs), they are relatively easy to PCR amplify from even degraded samples, such as from feces (Barbosa et al. 2013) or museum skins (Neumann and Jansman 2004, Jones et al. 2005, McDonough et al. 2018).

Although generally assumed to be neutral (i.e., mutations do not have adaptive or evolutionary consequences), occasionally microsatellites have been identified as having causative roles in diseases, and are particularly prevalent among microsatellites with trinucleotide (e.g., AGA_n) repeats (Bagshaw 2017). In voles (*Microtus* spp.), size polymorphism at a microsatellite closely linked to vasopressin receptor gene (Avprla) contributes to gene expression and may be linked to monogamous mating patterns and social behavior (Young and Hammock 2007). Thus, one should not assume microsatellite markers are strictly neutral without first examining patterns of polymorphism closely. Tests for neutrality, such as the Ewens-Watterson test (Ewens 1972, Watterson 1978) can be implemented in programs such as PyPop (Lancaster et al. 2003, 2007), and tests for outlier loci are available in the R package DetSel (Vitalis et al. 2003) and in the program Arlequin (Excoffier and Lischer 2010). We recommend looking at Selkoe and Toonen (2006) as a starting point on why, how, and when to consider using microsatellite markers, keeping in mind that identifying and developing microsatellites markers for novel species is now readily done using next-generation sequencing and can be accomplished in a short period of time. For example, a set of 21 microsatellite markers for salt marsh voles (*Microtus pennsylvanicus*) were developed from long sequence reads (PacBio RSII), optimized, and screened in approximately four weeks (Austin et al. 2014). However, the use of microsatellites has begun to wane due to the ease of obtaining large numbers of single nucleotide polymorphisms (SNPs) from next-generation sequencing platforms. SNP studies are increasingly cost-effective on a per-marker basis and are now widely applied to studies of ecological, evolutionary, and applied genetics.

Single Nucleotide Polymorphisms (SNPs)

SNPs are the principal form of DNA variation in any animal or plant species, with over 150 million SNPs identified from the human genome (https://www.ncbi.nlm.nih.gov/SNP/). By definition, a SNP is a polymorphic nucleotide position in a DNA sequence. By their nature, individual SNPs have low polymorphism; there are only four possible states (A, C, T, G), and the likelihood of multiple point mutations at a nucleotide within a population or species is low; as a result, most SNPs are characterized as biallelic (i.e., having two allelic states). Although this low polymorphism makes it less informative per locus relative to microsatellites, SNPs can now be readily genotyped at thousands of loci simultaneously at a much lower cost per locus than microsatellites. They are also characterized as being more mutationally stable in mammals, reducing the chance of homoplasy (Markovtsova et al. 2000, Nielsen 2000). Their high abundance in the genome has made them useful for identifying quantitative trait loci (QTL) and the association of genes with specific ecologically important traits. For example, experimental hybridization between *Peromyscus* species led to the identification of SNPs associated with differences in burrowing behavior between *P. polionotus* and *P. maniculatus* (Metz et al. 2017).

The increased need to genotype many thousands of SNPs on large numbers of individuals (particularly in medical, agricultural, and aquaculture studies) has driven the development of SNP genotyping platforms. For example, SNP arrays, or chips, have been developed (e.g., Matsuzaki et al. 2004, Steemers and Gunderson 2007) for nonhuman species such as cattle, dog, pig, and sheep (https://www.thermofisher.com/us/en/home/life-science/micro array-analysis.html, https://www.illumina.com/science/technology/beadarray -technology.html). However, these SNP-chip technologies remain cost prohibitive to develop for many ecologists or empirical evolutionary biologists, and they cannot be easily applied to varying experimental design needs. In response, methods for generating SNP genotypes have advanced rapidly, with novel methods for discovering and genotyping SNPs in nonmodel species appearing every year, claiming improvements in efficiency and cost. In general, these methods are reliant upon next-generation sequencing (NGS) methods that reduce genome complexity (i.e., genome reduction), allowing for greater sample-processing ability. Below we provide an overview of the general methods now widely used for reducing genome complexity. It should be noted that this section is written as an introduction to some of the leading genomic approaches that are beneficial for evaluating large numbers of genetic markers scattered across the genome, not an exhaustive survey of genomic methods.

Genomic Applications for Studying Small Mammals

Advancements in genome sequencing technologies are moving research on small mammals forward in a number of ways. One way is that genome-scale data provide an ability to describe and evaluate both neutral and nonneutral (i.e., markers affecting fitness) regions of the genome, and thus providing insight into potentially adaptive genetic variation (e.g., adaptation to different environments) (Holderegger et al. 2006, Kohn et al. 2006, Allendorf et al. 2010). High-throughput NGS can also improve the ability of genetic approaches that focus on a small number (e.g., hundreds or thousands) of loci, rather than the whole genome. For example, NGS can be used to identify ecological communities from DNA in a single environmental sample (e.g., water, soil, or feces; see Chapter 2). Similarly, studies utilizing microsatellites can be conducted using high-throughput sequencing even on low-quality DNA (e.g., De Barba et al. 2017), relieving the need for maintaining increasingly antiquated capillary genotyping equipment. Related to this, the massive amount of sequence data generated by a single run on an NGS instrument allows for substantial proportions of the genome to be sampled faster and at a lower cost than ever before. In many cases, this increase in the number of loci

THE RAPIDLY ADVANCING FIELD OF DNA SEQUENCING

First-Generation Sequencing

Sanger's dideoxy sequencing (Sanger et al. 1977) was the first method to be used routinely for sequencing of DNA in the laboratory. The Sanger method involved a process referred to as "chain termination" followed by separation of fragments to reconstruct the order of dinucleotides. The advent of PCR allowed for the targeting of specific loci, which together with chain-termination sequencing revolutionized how we study genetics. First-generation sequencing required separate steps for PCR amplification, sequencing, and separation (by electrophoresis), which meant that automating the process was difficult, limiting throughput, scalability, and resolution. Applied Biosystems Inc. (ABI) produced the first widely used fluorescence-based sequencer that greatly improved throughput of chain-termination sequencing (and could be used for fragment analysis, e.g., microsatellite genotyping). The cost per base is high relative to modern technologies (Mardis 2011). However, Sanger's method remained one of the most precise sequencing methods, having 99.99% accuracy. This is now being matched by third-generation technologies.

Next-Generation Sequencing

Next-generation sequencing (NGS), massively parallel sequencing, or deep sequencing each describe the post-Sanger DNA sequencing technology that can be used to sequence entire genomes or to target specific areas of interest (Goodwin et al. 2016). The numerous NGS technologies allow millions of small (hundreds of base pairs, or bp) fragments to be produced in parallel, allowing an entire mammal genome to be sequenced in a day or two. By contrast, Sanger sequencing took more than 10 years to produce the final-draft human genome.

Second-generation sequencing platforms are classified into sequencing by ligation (e.g., SOLiD, Complete Genomics) or sequencing by synthesis (e.g., Illumina, 454, Ion Torrent) categories. These "short-read" platforms (generally producing sequence reads of hundreds of bases) vary in their relative accuracy and precision, with some platforms persisting and evolving (e.g., Illumina), while others have fallen behind in their application (e.g., 454) or disappeared altogether.

Third-generation sequencing platforms ("long-read" platforms) are pushing genome science forward by simplifying our understanding of complex portions of the genome. Because large portions of mammalian genomes consist of complex repeat regions, long reads (several kilobases) allow for easier resolution of these genome features (Huddleston et al. 2014). In addition, the study of transcriptomes is improved by allowing researchers to identify gene isoforms (i.e., the manner in which gene exons are spliced and connected to produce functional transcripts). Single-molecule real-time (SMRT) sequencing (e.g., PacBio, Oxford Nanopore) does not require clonal amplification. PacBio utilizes a fixed polymerase inside each of thousands of picoliter wells on a plate where a single DNA molecule 10,000 nucleotides long is read. The MinION from Nanopore is a leader in "portable" field-ready sequencers, and it is rapidly becoming a reality, at least for some applications (e.g., DNA barcoding) (Menegon et al. 2017).

The main factors associated with evaluating NGS technology are (1) throughput, (2) read length, (3) read accuracy, (4) read depth (number of times each base is sequenced in independent events), and (5) cost per base (Morey et al. 2013). As the price per base drops and accuracy, read length, and depth increase, the main limitation for researchers is big data management and analysis (i.e., bioinformatics) (Pettersson et al. 2009).

Being able to scale up had major implications for how research is done. With first-generation sequencing, research questions were beyond what could be addressed by the technology. Now, NGS technology is quickly giving rise to new research questions (Morey et al. 2013). For example, detailed information on the diet of rare species or invasive mammals can be illuminated by scat analysis (Zarzoso-Lacoste et al. 2016, Biffi et al. 2017). Information on the timing and extent of hybridization among mammal species has increased (Wall et al. 2016, Marques et al. 2017). Ecological adaptations are increasingly coming to light with NGS tools, for example, fine-scale ecological adaptation to soil conditions in *Spalax galili* (blind mole-rats) (Li et al. 2015) and adaptation to urban environments in *Peromyscus leucopus* (white-footed mouse) (Harris and Munshi-South 2017). Physiological adaptations at the molecular level have also been uncovered in a variety of mammal species (see, for example, Velotta et al. 2016, Kordonowy and MacManes 2017, Giorello et al. 2018). A good review of NGS technologies and their application in mammals is found in Larsen and Matocq (2019).

Table 15.1. Examples of ecological and evolutionary research questions that can be addressed through genome reduction approaches

Research topic	Approach	Application
Adaptation to different environments	RNA-Seq	Transcriptional analyses identified how precise mutations and the interactions of different mutations affect aerobic performance in deer mice (*Peromyscus maniculatus*) across altitudinal gradients (Storz et al. 2009, Natarajan et al. 2013)
Adaptation to different environments	RNA-Seq	Plasticity in gene regulation and expression among high and low elevation deer mice (*P. maniculatus*) explains acclimation to hypoxic cold stress (Velotta et al. 2016)
Physiological adaptation	RNA-Seq	Discovery of genes associated with metabolism and immune function showing signatures of selection in urban populations of white-footed mice *Peromyscus leucopus* (Harris and Munshi-South 2017)
Adaptation to climate change	Exome sequencing	Directional selection detected in alpine chipmunks (*Tamias alpinus*) in response to climate warming and range contraction over past century (Bi et al. 2013)
Adaptation to different environments	RNA-Seq	Evidence for selection at renal transcripts in the desert-adapted cactus mice (*P. eremicus*) (MacManes and Eisen 2014)
Behavior	RADseq	Identification of genomic regions associated with complex burrowing behavior in deer mice (Weber et al. 2013)
Conservation and management	RADseq	Assessment of genetic consequences of reintroduction programs involving multiple small mammal species (White et al. 2018)
Cryptic species identification/ detecting hybridization	RADseq	Identification of 117 diagnostic SNP loci between yellow-necked mice (*Apodemus flavicollis*) and wood mice (*A. sylvaticus*) for studies on hybridization and introgression in natural populations (Cerezo et al. 2019)
Deep phylogenetic relationships	Target enrichment	Libraries of enriched areas of ultraconserved elements (UCEs) greatly increase the resolution of contentious relationships including the monophyly of the Glires clade (rodents and lagomorphs) and the basal position of the hedgehogs (Erinaceinae) within the Laurasiatheria (McCormack et al. 2012)
Post-Pleistocene phylogeography	RNA-Seq	Intraspecific genome replacement detected by invading populations of bank voles (*Clethrionomys glareolus*) in modern Great Britain during the Pleistocene-Holocene transition (Kotlik et al. 2018)
Population connectivity	RADseq	Kinship estimated from SNPs demonstrated the impact of watershed topography, but not rivers, on dispersal in Pyrenean desman (*Galemys pyrenaicus*) (Escoda et al. 2018)
Responses to selection	RNA-Seq	Gene expression changes in bank voles (*Myodes glareolus*) selected for increased aerobic metabolism relative to controls (Konczal et al. 2015)
Wildlife disease	Exome sequencing	Quantified low functional diversity in immune genes in Tasmanian devils *Sarcophilus harrisii* (Morris et al. 2015)
Demographic disruption following natural disasters	Target enrichment	Demonstrated the genomic impact of volcanic activity on two species of tuco-tuco *Ctenomys sociabilis* and *C. haigi* (Hsu et al. 2017)

(Munshi-South et al. 2016), and delineating cryptic (morphologically indistinguishable) lineages of fossorial small mammals (Mynhardt et al. 2020).

Target Enrichment

Target enrichment involves selectively sampling a genomic library using hybridization techniques or biotinylated probes that are complementary to specific genomic regions. Probes bind to the target region and the captured DNA fragments are then isolated from the remaining solution (which may consist of exogenous DNA, for example). These captured (enriched) regions are typically PCR amplified to increase their copy number and then sequenced. However, unlike RADseq approaches that typically identify anonymous SNPs, target capture methods enrich for specific sites, like exons (see exome sequencing, below) or introns, or other subsets of the genome of interest. For example, target enrichment for the mitochondrial genome of Sunda colugo (*Galeopterus variegatus*) museum specimens recovered 50% to 94% of the total mitochondrial genome across 13 individual skins ranging in age from 47 to 170 years (Mason et al. 2011). Targeted enrichment permits sequencing at greater depth of coverage than whole-genome sequencing. One advantage of this greater depth of coverage is that it increases the chance of discovering rare gene variants (see RNA-seq, below). While this approach is still predominantly used to study human disease, the methods can be transferred to relevant questions on small mammals and other nonmodel organisms. For example, ultraconserved elements (UCEs) are commonly targeted for resolving very old evolutionary relationships in phylogenetics. Targeted sequencing of immune genes in the larger Tasmanian devil (*Sarcophilus harrisii*) has provided critical information about the low levels of diversity in the remaining population (Morris et al. 2015). Sequence-capture methods coupled with NGS provide high-coverage sequence data from multiple individuals or populations, making this approach desirable to molecular ecologists (Andrews and Luikart 2014).

Exome sequencing is a specific application of target enrichment. The exome consists of all the DNA sequences representing all the exons of protein-coding genes and covers ~2% of mammal genomes. This makes exome capture a cost-effective strategy for studying genome-wide adaptation. Exome capture requires knowledge of the organism's genome, so an initial reference genome project, or a reference genome from a closely related species, is required to design "baits," which are oligonucleotide probes (each 60–120 nucleotides long) that hybridize to the gene or gene regions of interest (Mamanova et al. 2010, Andrews and Luikart 2014). Exome capture is also useful for recovering target DNA from low-quality samples rich in exogenous DNA (e.g., feces, bone, or museum samples). An exon capture approach was used to sequence alpine chipmunk (*Tamias alpinus*) museum skins to exam-ine changes in diversity coincident with climate-induced range contraction in the Sierra Nevada mountain range in North America (Bi et al. 2013). There are various commercial platforms developed for exome sequencing, and a good overview can be found in Warr et al. (2015).

The development of probes requires having a good-quality reference ge-nome for the target species. Closely related species often will work; however, this may be affected by the degree of synteny among closely related species. More recent advancements have been able to target capture without reference genomes, although the widespread application of such methods remains lim-ited for ecological and evolutionary research questions on nonmodel species, including small mammals (Jones and Good 2016).

RNA Sequencing

Genome complexity can also be reduced by focusing on the process of tran-scription (see Table 15.2 for distinction between exome sequencing and RNA sequencing). RNA sequencing (RNA-Seq) focuses on the RNA isolated from tissues that is then reverse transcribed to produce double stranded comple-mentary DNA (cDNA) for sequencing using commercially available kits. As for other genome-reduction approaches, multiple individuals can be pooled for sequencing, allowing for transcriptional variation and SNP discovery within expressed portions of the genome.

Reducing analyses to the transcribed portions affords a number of po-tential advantages. First, as mentioned, by reducing the complexity of the ge-nome, one can obtain information on expressed genes from many individuals sequenced simultaneously. A second advantage specific to RNA-Seq is that, because SNPs identified from RNA-Seq are located within expressed genes

Table 15.2. Comparison of exome sequencing and RNA sequencing

	Exome	Transcriptome
Approach	Exome sequencing	RNA sequencing
Description	~2% of the genome consists of exons, which are transcribed and translated during protein synthesis	The total of all messenger RNA molecules that is transcribed from exons
Made up of	DNA	RNA
Consists of	Total set of exons in genome	Total set of mRNA expressed in a tissue type
Application	Variation in exon sequence possibly due to natural selection; longitudinal studies of populations, habitat associations	Detection of transcriptional variation among sex, age, habitats; studies of transcriptional response to wildlife diseases

rather than anonymous genomic regions, RADseq can aid in the identification of genes if functional characterization of observed variation is of interest. Furthermore, because the number of sequences obtained will be proportionate to the amount of different transcripts in the sampled tissue (unless library normalization is conducted prior to sequencing; e.g., Christodoulou et al. 2011), RNA-Seq can, for example, provide quantitative information for gene expression analysis in different environments, such as white-footed mice (*Peromyscus leucopus*) across the rural-urban gradient (e.g., Harris et al. 2013, Harris and Munshi-South 2017).

The quality of resulting SNP data generated from transcripts will be greatly improved by aligning sequencing reads to a reference genome or reference transcriptome of a conspecific or closely related species (De Wit et al. 2015). Thus, RNA-Seq is often seen as a first step for research focused on studying transcriptome variation related to ecology or behavior (e.g., MacManes and Lacey 2012). By having an annotated reference transcriptome, SNPs identified from RNA-Seq can be identified to gene regions, unlike anonymous SNPs generated from RADseq.

Disadvantages of RNA-Seq with respect to ecological or evolutionary studies on small mammals include issues with RNA preservation in the field. First, like DNA, RNA quality is critical for RNA-Seq, but unlike DNA, RNA is a single-stranded molecule and thus high unstable. Commercially available buffers for preserving RNA are available (RNA*later*, Thermo Fisher; DNA/RNA Shield, Zymo Research); however, they should not be assumed to be suitable for maintaining RNA stability for long periods in the field without refrigeration. In addition, environmental contamination from exogenous RNA requires careful bioinformatic filtering (e.g., Chu et al. 2014).

RNA-Seq has yet to become a widespread approach to obtaining reduced representation libraries for population genetic studies of small mammals for reasons mentioned above and because of the latency of translating from medical research to molecular ecology. Outside of comparative expression studies (Sudmant et al. 2015) or medical studies utilizing model mammal species (e.g., Mortazavi et al. 2008), RNA-Seq has been applied to characterize the study of monotreme venom (Wong et al. 2013) and to study patterns of protein evolution in wild European rabbits (*Oryctolagus cuniculus*) (Carneiro et al. 2012) and RNA performance during hibernation in ground squirrels (Anderson et al. 2016).

Further Considerations

It is frequently emphasized that genome-reduction approaches "simplify" genome typing (Luikart et al. 2003). While true, the quality of the output is very sensitive to DNA quality input, bioinformatics filtering decisions, and exper-

common (Jane et al. 2015) and their impact on assays should be tested during pilot phases of eDNA studies. Pilot studies should also be conducted to determine what combination of sampling, replication, and extraction methodology is most suitable to maximize eDNA yield and increase detection probability for individual studies. A good example of how laboratory protocols affect eDNA detection is given by Piggott (2016).

Population Genetics of Small Mammals

Population genetics aims to understand the evolutionary forces that affect the frequency and distribution of alleles within and among populations. These forces include genetic drift (random changes in allele frequencies due to small breeding population sizes), natural selection, mutations, genetic recombination, and the movement of alleles between populations (gene flow). Small mammal populations vary widely in their sizes, dynamics, and genetic structuring. Population structure (how populations are organized) is also widely variable across species. Single populations that are isolated from others, with all their reproductive individuals having the opportunity to mate with one another, are characterized as being panmictic. Alternatively, populations of small mammals can be highly subdivided, where individuals within one subpopulation are far more likely to mate with each other than with individuals in other subpopulations. Metapopulations reflect a series of these subpopulations, or demes, that are characterized by periodic extinctions and recolonizations (Hanski 1998). Numerous small mammals exhibit metapopulation structure, particularly wetland-associated species such as marsh rice rats, *Oryzomys palustris* (van der Merwe et al. 2016), and round-tailed muskrats, *Neofiber alleni* (Schooley and Branch 2009). The extent to which small mammal populations will be segregated into demes ranges from scarcely at all to situations where movement (and subsequent reproduction) hardly happens.

Population Structure

Population structuring will also be determined by the geographic scale of the study. Even wide-ranging species that have limited deme structure at finer spatial scales will be differentiated to some degree across their biogeographic range due to natural barriers, such as rivers and mountains, or due to anthropogenic barriers, including habitat fragmentation. Many small mammals are also highly dynamic, changing in density and distribution over time, which also has important genetic consequences. Deciphering the impacts of these population dynamics on the models described above is a central tenet of molecular ecology.

Population connectivity (gene flow) determines the dynamics and evolution of populations, making it a critical component of ecology and conservation biology. Similarly, studying populations requires an understanding of their extent and delimitations. Many of the markers discussed above (e.g., microsatellites and SNPs) have been instrumental in assessing population genetic questions. However, the genome-reduction methods introduced here allow us to go beyond these traditional population genetic parameters. For example, with genome-scale representation, outlier loci having greater divergence than the background levels may be indicative of natural selection acting on this loci. Similarly, exome or RNA sequencing can be used to compare gene variation or differences in expression among populations inhabiting different habitat types. Knowing specific markers associated with traits of interest may be of value to ecologists interested in, for example, understanding mechanisms leading to boldness (Chapter 13) in different habitats, or to study the evolutionary response to rodenticides (Rost et al. 2009).

Measuring Genetic Diversity

In general, genetic diversity of populations is measured using three main metrics: heterozygosity, allelic richness, and proportion of polymorphic loci. These measures have typically been applied to genetic markers that are predicted to be neutral with respect to natural selection. At neutral markers, genetic differences between populations will arise from genetic drift and from mutations. Because most of the genome is expected to be neutral (or nearly so), this idea can often be safely assumed, particularly when looking at markers from outside the exome. Another major assumption when seeking to quantify genetic diversity is that neutral variation is correlated (if only weakly) with adaptive variation (Reed and Frankham 2001, 2002, Holderegger et al. 2006). If true, this assumption would allow us to infer that the amount of neutral genetic variation reflects the adaptive potential of a population. Furthermore, population genetic analyses assume that the genotype frequencies can be estimated from allele frequencies (see "The Hardy-Weinberg Principle"). This expectation relies on a number of assumptions that can be rather restrictive when applied to wild populations. However, they are useful in that deviations from the Hardy-Weinberg equilibrium can be informative about population attributes of interests, such as hidden population structure and nonrandom mating within populations (e.g., inbreeding). When examining multiple markers, tests for linkage disequilibrium (LD), or the nonrandom association of alleles between loci, are commonly applied. This is because many population genetic analyses (e.g., estimating the number of genetic populations represented in your dataset) make assumptions about the independence of loci. Linkage can occur when loci are physically linked when they occur on the same chromosome. Recombination will breakdown linkage

the thousands) with lower polymorphism does present analytical challenges (specifically in terms of data processing). Those starting out on population genetics projects are encouraged to read Meirmans (2015) for advice on planning sampling strategies.

Effective population size. In population ecology, the census size (N) is estimated as either the total number of individuals, the total number of adults, or the total number of breeding adults (Nunney and Elam 1994). However, in genetic terms, the population size is typically expressed in terms of the effective population size (N_e), which was originally defined (Fisher 1930, Wright 1931) as the number of individuals in an idealized population, characterized by having random mating, equal sex ratios, and no immigration or emigration. In addition, genetic mutations and the effect of natural selection are assumed to be negligible. The advantage of this seemingly strict definition is that it allows for a standard measure of population size that can then be used to compare populations having very different ecological characteristics or life histories, but it also allows us to understand the genetic consequences of variation in census population sizes (Rowe and Lougheed 2019). N_e is required for predicting the rate at which genetic variation will be lost within a population, making it a critical conservation genetic parameter (Luikart et al. 2010). For small mammals, the ratio of N_e/N is predicted to be low due to some typical characteristics of the life history of small mammals. For example, N_e is expected to be lower in populations that vary in census size over time, typical in many small mammals. In addition, small mammals with skewed sex ratios and high variance in reproductive output will possess lower N_e than species or populations with balanced sex ratios or small variance in reproductive output. For example, variation in litter sizes among conservation breeding colonies of common hamsters (*Cricetus cricetus*) was positively associated with estimates of N_e in reintroduced populations sourced from different colonies (La Haye et al. 2017).

Sampling requirements for estimating N_e vary depending on the analytical method applied to estimate the parameter. For example, there are one-sample (sampled contemporaneously), multisample, or temporal methods, each providing different strengths and limitations. Within each, different approaches have been developed that utilize different genetic parameters (e.g., LD, heterozygote excess) as the basis of the estimate of N_e. These approaches have been the focus of reviews and thus are not detailed here. See Charlesworth (2009) and Luikart et al. (2010) for detailed overviews of the underlying theory and assumptions of estimating N_e, and see Wang et al. (2016) for a recent overview of the main developments for predicting N_e. See Rowe and Lougheed (2019) for a recent list of software programs that can be used for estimating N_e and N from genetic samples.

As discussed in Chapter 9, census population size (N) is one of the fundamental properties of organisms that is of interest to ecologists. However, geneticists are also interested in the effective population size (N_e), which reflects the size of an idealized population that is experiencing the same rate of genetic change as that of the natural population under consideration (Crow and Kimura 1970). Despite the unrealistic assumptions associated with an idealized population (random mating, equal family sizes, and free from the effects of selection, migration, or mutation), estimating N_e allows us to compare across populations and species with different ecologies, and it provides insight into the genetic consequences of different population sizes. There are many ways of estimating N_e, which are beyond the purview of this chapter, but one important attribute is whether the estimator incorporates historical or contemporary demographic change.

Conclusion and Future Directions

This chapter provides a cursory look at molecular ecological techniques as applied to small mammals. Because of the nature of the study of genetics, much of what can be applied to other organisms (particularly vertebrates) applies to small mammals as well. As such, we encourage those interested in applying molecular techniques to study ecological or evolutionary questions related to small mammals to explore the literature. While genetic papers do appear in taxon-specific mammalogy journals, there are a number of top-ranked publications (e.g., *Molecular Ecology, Ecological Applications, Conservation Genetics*) to explore that will provide insight into the types of exciting research that can be applied to your study system.

Improvements in marker development and genome representation are advancing important fields of study not focused on in this book. For example, phylogeography is a field that focuses on understanding the forces that determine the geographic distribution of genetic lineages (Avise 2000) and has been a major field of study with respect to mammals. Initially emphasizing relatively descriptive studies limited to mitochondrial genes, phylogeography has blossomed into a model-based field (Beaumont et al. 2010) capable of testing support for alternative hypotheses. The advancement of phylogeography with the incorporation of NGS approaches and historical specimens (McLean et al. 2016) will rapidly advance our understanding of the evolutionary histories of small mammals. Along with population genetics and the field of landscape genetics, the application of NGS will grow in its value in providing greater precision to our insights on factors affecting patterns of gene flow, with implications for our understanding of species ecology and for management. For example, RADseq tools have provided insight into the ge-

REFERENCES

Aars, J., and R. A. Ims 1999. "The effect of habitat corridors on rates of transfer and interbreeding between vole demes." *Ecology,* 80: 1648–1655.

Adler, G. H., and R. Levins. 1994. "The island syndrome in rodent populations." *Quarterly Review of Biology* 69: 473–490.

Allendorf, F. W., P. A. Hohenlobe, and G. Luikart. 2010. "Genomics and the future of conservation genetics." *Nature Reviews Genetics* 11: 697–709.

Anderson, K. J., K. L. Vermillion, P. Jagtap, J. E. Johnson, T. J. Griffin, and M. T. Andrews. 2016. "Proteogenomic analysis of a hibernating mammal indicates contribution of skeletal muscle physiology to the hibernation phenotype." *Journal of Proteomic Research* 15: 1253–1261.

Andrews, K. R., J. M. Good, M. R. Miller, G. Luikart, and P. A. Hohenlohe. 2016. "Harnessing the power of RADseq to ecological and evolutionary genomics." *Nature Reviews Genetics* 17: 81–92.

Andrews, K. R., and G. Luikart. 2014. Recent novel approaches for population genomics data analysis. *Molecular Ecology* 23: 1661–1667.

Angley, L. P., M. Combs, C. Firth, M. J. Frye, W. I. Lipkin, J. L. Richardson, and J. Munshi-South. 2018. "Spatial variation in the parasite communities and genomic structure of urban rats in New York City." *Zoonoses and Public Health* 65: e113–e123.

Aplin K. P., H. Suzuki, A. A. Chinen, R. T. Chesser, J. ten Have, S. C. Donnellan, et al. 2011. "Multiple geographic origins of commensalism and complex dispersal history of black rats." *PLoS ONE* 6: e26357.

Arakawa, Y., C. Nishida-Umehara, Y. Matsuda, S. Sutos, and H. Suzuki. 2002. "X-chromosomal localization of mammalian Y-linked genes in two XO species of the Ryukyu spiny rat." *Cytogenetics and Genome Research* 99: 303–309.

Austin, J. D., D. U. Greene, R. L. Honeycutt, and R. A. McCleery. 2018. "Genetic evidence indicates ecological divergence rather than geographic barriers structure Florida fox squirrels." *Journal of Mammalogy* 99: 1336–1349.

Austin, J. D., J. A. Gore, D. U. Greene, and C. Gotteland. 2015. "Conspicuous genetic structure belies recent dispersal in an endangered beach mouse (*Peromyscus polionotus trissyllepsis*)." *Conservation Genetics* 16: 915–928.

Austin, J. D., E. V. Saarinen, A. Arias-Perez, R. A. McCleery, and R. H. Lyons. 2014. "Twenty-one novel microsatellite loci for the endangered Florida salt marsh vole *Microtus pennsylvanicus dukecampbelli*)." *Conservation Genetics Resources* 6: 637–639.

Avise, J. C. 2000. *Phylogeography: The History and Formation of Species.* Cambridge: Harvard University Press.

Bagshaw, A. T. M. 2017. "Functional mechanisms of microsatellite DNA in Eukaryotic genomes." *Genome Biology and Evolution* 9: 2428–2443.

Baird, N. A., P. D. Etter, T. S. Atwood, M. C. Currey, A. Shiver, Z. A. Lewis, et al. 2008. "Rapid SNP discovery and genetic mapping using sequenced RAD markers." *PLoS ONE* 3: e3376.

Baker, R. J., and R. D. Bradley. 2006. "Speciation in mammals and the genetic species concept." *Journal of Mammalogy* 87: 643–662.

Ballard, J. W. O., and D. M. Rand. 2005. "The population biology of mitochondrial DNA and its phylogenetic implications." *Annual Review of Ecology Evolution and Systematics* 36: 621–642.

Ballard, J. W. O., and M. C. Whitlock. 2004. "The incomplete natural history of the mitochondria." *Molecular Ecology* 13: 729–744.

Barbosa, S., J, Pauperio, J. B. Searle, and P. C. Alves. 2013. "Genetic identification of Iberian rodent species using mitochondrial and nuclear loci: Application to noninvasive sampling." *Molecular Ecology Resources* 13: 43–56.

Beaumont M. A., R. Nielsen, C. Robert, J. Hey, O. Gaggiotti, L. Knowles, et al. 2010. "In defence of model-based inference in phylogeography." *Molecular Ecology* 19 (3): 436–446.

Benestan, L. M., A.-L. Ferchaud, P. A. Hohenlohe, B. A. Garner, G. J. P. Naylor, I. B. Baums, et al. 2016. "Conservation genomics of natural and managed populations: Building a conceptual and practical framework." *Molecular Ecology* 25: 2967–2977.

Bi, K., T. Linderoth, D. Vanderpool, J. M. Good, R. Neilson, and C. Moritz. 2013. "Unlocking the vault: Next generation museum population genomics." *Molecular Ecology* 22: 6018–6032.

Biffi, M., F. Gillet, P. Laffaille, F. Colas, S. Aulagnier, F. Blanc, et al. 2017. "Novel insights into the diet of the Pyrenean desman (*Galemys pyrenaicus*) using next-generation sequencing molecular analyses." *Journal of Mammalogy* 98 (5): 1497–1507. https://doi.org/10.1093/jmammal/gyx070.

Brace, S., E. Palkopoulou, L. Dalén, A. M. Lister, R. Miller, M. Otte, et al. 2012. "Serial population extinctions in a small mammal indicate Late Pleistocene ecosystem instability." *Proceedings of the National Academy of Sciences USA* 109 (50): 20532–20536. https://doi.org/10.1073/pnas.1213322109.

Bryja, J., and A. Konecny 2003. "Fast sex identification in wild mammals using PCR amplification of the *Sry* gene." *Folia Zoologica* 52 (3): 269–274.

Carneiro, M., F. W. Albert, J. Melo-Ferreira, N. Galtier, P. Gayral, J. A. Blanco-Aguiar, et al. 2012. "Evidence for widespread positive and purifying selection across the European Rabbit (*Oryctolagus cuniculus*) genome." *Molecular Biology and Evolution* 29: 1837–1849. https://doi.org/10.1093/molbev/mss025.

Cerezo, M. L. M., M. Kucka, K. Zub, Y. F. Chan, and J. Bryk. 2019. "Population structure of *Apodemus flavicollis* and comparison to *Apodemus sylvaticus* in northern Poland based on whole-genome genotyping with RAD-seq." *bioRxiv*. https://doi.org/10.1101/625848.

Chalmel, F., A. D. Rolland, C. Niederhauser-Wiederkehr, S. S. W. Chung, P. Demougin, A. Gattiker, et al. 2007. "The conserved transcriptome in human and rodent male gametogenesis." *Proceedings of the National Academy of Sciences, USA* 104: 8346–8351.

Charlesworth, B. 2009. "Effective population size and patterns of molecular evolution and variation." *Nature Reviews Genetics* 10: 195–205.

Christodoulou, D. C., J. M. Gorham, D. S. Herman, and J. G. Seidman. 2011. "Construction of normalized RNA-seq libraries for next-generation sequencing using the crab duplex-specific nuclease." *Current Protocols in Molecular Biology* 94: 4.12.1–4.12.11.

Chu, J., S. Sadeghi, A. Raymond, K. M. Nip, R. Mar, H. Mohamadi, et al. 2014. "BioBloom tools: Fast accurate and memory-efficient host species sequence screening using bloom filters." *Bioinformatics* 30 (23): 3402–3404.

et al. 2015. "Distance, flow and PCR inhibition: eDNA dynamics in two headwater streams." *Molecular Ecology Resources* 15 (1): 216–227.

Jeffreys, A. J., V. Wilson, and S. L. Thein. 1985. "Hypervariable 'minisatellite' regions in human DNA." *Nature* 314: 67–73.

Jones, M. R., and J. M. Good. 2016. "Targeted capture in evolutionary and ecological genomics." *Molecular Ecology* 25: 185–202.

Jones, R. T., A. P. Martin, A. J. Mitchell, S. K. Collinge, and C. Ray. 2005. "Characterization of 14 polymorphic microsatellite markers for the black-tailed prairie dog (*Cynomys ludovicianus*)." *Molecular Ecology Resources* 5: 71–73.

Kim, E. B., X. Fang, A. A. Fushan, Z. Huang, A. V. Lobanov, L. Han, et al. 2011. "Genome sequencing reveals insights into physiology and longevity of the naked mole rat." *Nature* 479: 223–227. https://doi.org/10.1038/nature10533.

Koepfli, K.-P., B. Paten, the Genome 10K Community of Scientists, and S. J. O'Brien. 2015. "The Genome 10K Project: A way forward." *Annual Reviews of Animal Biosciences* 3: 57–111.

Kohn, M. H., W. J. Murphy, E. A. Ostrander, and R. K. Wayne. 2006. "Genomics and conservation genetics." *Trends in Ecology and Evolution* 21: 629–637.

Kohn, M. H., and R. K. Wayne. 1997. "Facts from feces revisited." *Trends in Ecology and Evolution* 12: 223–227.

Konczal, M., W. Babik, J. Radwan, E. T. Sadowska, and P. Koteja. 2015 "Initial molecular-level response to artificial selection for increased aerobic metabolism occurs primarily through changes in gene expression." *Molecular Biology and Evolution* 32 (6): 1461–1473. https://doi.org/10.1093/molbev/msv038.

Kordonowy, L., and M. MacManes. 2017. "Characterizing the reproductive transcriptomic correlates of acute dehydration in males in the desert-adapted rodent, *Peromyscus eremicus*." *BMC Genomics* 18: 473.

Kotlik, P., S. Marková, M. Konczal, W. Babik, and J. B. Searle. 2018. "Genomics of end-Pleistocene population replacement in a small mammal." *Proceedings of the Royal Society B.* 285 (1872): 2017.2624. https://doi.org/10.1098/rspb.2017.2624.

Kozakiewicz, M., A. Choluj, A. Kozakiewicz, and M. Sokó. 2009. "Familiarity and female choice in the bank vole—do females prefer strangers?" *Acta Theriologica* 54: 157–164.

Kuroiwa, A., S. Handa, C. Nishiyama, E. Chiba, F. Yamada, S. Abe, and Y. Matsuda. 2011. "Additional copies of *CBX2* in the genomes of males of mammals lacking *SRY*, the Amami spiny rat (*Tokudaia osimensis*) and the Tokunoshima spiny rat (*Tokudaia tokunoshimensis*)." *Chromosome Research* 1: 635–644.

La Haye, M. J. J., T. E. Reiners, R. Raedts, V. Verbist, and H. P. Koelewijn. 2017. "Genetic monitoring to evaluate reintroduction attempts of a highly endangered rodent." *Conservation Genetics* 18: 877–892.

Lancaster, A. K., M. P. Nelson, D. Meyer, R. M. Single, and G. Thomson. 2003. "PyPop: A software framework for population genomics: Analyzing large-scale multi-locus genotype data." *Pacific Symposium on Biocomputing* 2003: 514–525.

Lancaster, A.K., R. M. Single, O. D. Solberg, M. P. Nelson, and G. Thomson. 2007. "PyPop update—a software pipeline for large-scale multilocus population genomics." *Tissue Antigens* 69: 192–197.

Larsen, P. A., and M. D. Matocq. 2019. "Emerging genomic applications in mammalian ecology, evolution, and conservation." *Journal of Mammalogy* 100 (3): 786–801. https://doi.org/10.1093/jmammal/gyy184.

Li, K., W. Hong, H. Jiao, G. D. Wang, K. A. Rodriguez, R. Buffenstein, et al. 2015. "Sympatric speciation revealed by genome-wide divergence in the blind mole rat *Spalax*." *Proceedings of the National Academy of Sciences, USA* 112: 11905–11910. https://doi.org/10.1073/pnas.1514896112.

Luikart G., P. R. England, D. Tallmon, S. Jordan, and P. Taberlet. 2003. "The power and promise of population genomics: From genotyping to genome typing." *Nature Reviews Genetics* 4: 981–994.

Luikart, G., N. Ryman, D. A. Tallmon, M. K. Schwartz, and F. W. Allendorf. 2010. "Estimation of census and effective population size: The increasing usefulness of DNA-based approaches." *Conservation Genetics* 11: 355–373. https://doi.org/10.1007/s10592-010-0050-7.

MacManes, M. D., and M. B. Eisen. 2014. "Characterization of the transcriptome, nucleotide sequence polymorphism, and natural selection in the desert adapted mouse *Peromyscus eremicus*." *PeerJ* 2: e642. https://doi.org/10.7717/peerj.642.

MacManes, M. D., and E. A. Lacey. 2012. "The social brain: Transcriptome assembly and characterization of the hippocampus from a social subterranean rodent, the colonial tuco-tuco (*Ctenomys sociabilis*)." *PLoS ONE* 9: e45524.

Maddison, W. P., and L. L. Knowles. 2006. "Inferring phylogeny despite incomplete lineage sorting." *Systematic Biology* 55: 21–30.

Mamanova, L., A. J. Coffey, C. E. Scott, I. Kozarewa, E. H. Turner, A. Kumar, et al. 2010. "Target-enrichment strategies for next-generation sequencing." *Nature Methods* 7: 111–118.

Mardis, E. R. 2008. "Next-generation DNA sequencing methods." *Annual Review of Genomics and Human Genetics* 9: 387–402.

Mardis, E. R. 2011. "A decade's perspective on DNA sequencing technology." *Nature* 470: 198–203.

Markovtsova, L., P. Marjoram, and S. Tavaré. 2000. "The age of a unique genetic polymorphism." *Genetics* 156: 401–409.

Marques, J. P., L. Farelo, J. Vilela, D. Vanderpool, P. C. Alves, J. M. Good, et al. 2017. "Range expansion underlies historical introgressive hybridization in the Iberian hare." *Scientific Reports* 7: 40788.

Mason, V. C., G. Li, K. M. Helgen, and W. J. Murphy. 2011. "Efficient cross-species capture hybridization and next-generation sequencing of mitochondrial genomes from noninvasively sampled museum specimens." *Genome Research* 21: 1695–1704. https://doi.org/10.1101/gr.120196.111.

Matsuzaki, H., S. Dong, H. Loi, E. Hubbell, J. Law, T. Berntsen, et al. 2004. "Genotyping over 100,000 SNPs on a pair of oligonucleotide arrays." *Nature Methods* 1: 109–111.

McCormack, J. E., B. C. Faircloth, and T. C. Glenn. 2012. "Ultraconserved elements are novel phylogenomic markers that resolve placental mammal phylogeny when combined with species-tree analysis." *Genome Research* 22 (4): 746–754.

McDonough, M. M., L. D. Parker, N. R. McInerne, M. G. Campana, and J. E. Maldonado. 2018. "Performance of commonly requested destructive museum samples for mammalian genomic studies." *Journal of Mammalogy* 99: 789–802.

McLean, B. S., K. C. Bell, J. L. Dunnum, B. Abrahamson, J. P. Colella, E. R. Deardorf,

Reuter, J. A., D. V. Spacek, and M. P. Snyder. 2015. "High-throughput sequencing technologies." *Molecular Cell Review* 58: 586–597.

Rokas, A., and P. Abbot. 2009. "Harnessing genomics for evolutionary insights." *Trends in Ecology and Evolution* 24: 192–200.

Rost, S., H.-J. Pelz, S. Menzel, A. D. MacNicoll, V. Leon, K.-J. Song, et al. 2009. "Novel mutations in the VKORC1 gene of wild rats and mice—a response to 50 years of selection pressure by warfarin?" *BMC Genetics* 10: 4. https://doi.org/10.1186/1471-2156-10-4.

Rousset, F., and M. Raymond. 1997. "Statistical analysis of population genetic data: New tools, old concepts." *Trends in Ecology and Evolution* 12: 313–317.

Rowe, G., M. Sweet, and T. Beebee. 2017. *An Introduction to Molecular Ecology*, 3rd ed. Oxford: Oxford University Press.

Rowe, J. R., and S. L. Lougheed. 2019. "Genetic insights into population ecology." In *Population Ecology in Practice*, edited by D. L. Murray and B. K. Sandercock, 263–298. Hoboken, NJ: John Wiley & Sons.

Saiki, R. K., S. Scharf, F. Faloona, K. B. Mullis, G. T. Horn, H. A. Erlich, et al. 1985. "Enzymatic amplification of beta-globin genomics sequences and restriction site analysis for diagnosis of sickle cell anemia." *Science* 230: 1350–1354.

Sanger, F., S. Nicklen, and A. R. Coulson. 1977. "DNA sequencing with chain-terminating inhibitors." *Proceedings of the National Academy of Sciences U.S.A.* 74: 5463–5467.

Schooley, R. L., and L. C. Branch. 2009. "Enhancing the area-isolation paradigm: Habitat heterogeneity and metapopulation dynamics of a rare wetland mammal." *Ecological Applications* 19 (7): 1708–1722. https://doi.org/10.1890/08-2169.1.

Scornavacca, C., K. Belkhir, J. Lopez, R. Dernat, F. Delsuc, E. J. P. Douzery, and V. Ranwez. 2019. "OrthoMaM v10: Scaling-up orthologous coding sequences and exon alignments with more than one hundred mammalian genomes." *Molecular Biology and Evolution* 36: 861–862. https://doi.org/10.1093/molbev/msz015.

Selkoe, K. A., and R. J. Toonen. 2006. "Microsatellites for ecologists: A practical guide to using and evaluating microsatellite markers." *Ecology Letters* 9: 615–629.

Shafer, A. B. A., J. B. W. Wolf, P. C. Alves, L. Bergsrom, M. W. Bruford, I. Brannstrom, et al. 2015. "Genomics and the challenging translation into conservation practice." *Trends in Ecology and Evolution* 30: 78–87.

Shaw, C. N., P. J. Wilson, and B. N. White. 2003. "A reliable molecular method of gender determination for mammals." *Journal of Mammalogy* 84: 123–128.

Sinclair, A. H., P. Berta, M. S. Palmer, J. R. Hawkins, B. L. Griffiths, M. J. Smith, et al. 1990. "A gene from the human sex-determining region encodes a protein with homology to a conserved DNA-binding motif." *Nature* 346: 240–244.

Song, S., L. Liu, S. V. Edwards, and S. Wu. 2012. "Resolving conflict in eutherian mammal phylogeny using phylogenomics and the multispecies coalescent model." *Proceedings National Academy of Sciences USA* 109: 14942–14947.

Springer, M. S., and J. Gatesy. 2016. "The gene tree delusion." *Molecular Phylogenetics and Evolution* 94: 1–33.

Springer, M. S., R. W. Meredith, E. C. Teeling, and W. J. Murphy. 2013. "Technical comment on 'The placental mammal ancestor and the post-K-Pg radiation of placentals.'" *Science* 341: 613.

Steemers, F. J., and K. L. Gunderson. 2007. "Whole genome genotyping technologies on the BeadArray platform." *Biotechnology Journal* 2: 41–49.

Steiner, C. C., A. S. Putnam, P. E. A. Hoeck, and O. A. Ryder. 2013. "Conservation genomics of threatened animal species." *Annual Review of Animal Bioscience* 1: 261–281.

Steppan, S. J., M. R. Akhverdyan, E. A. Lyapunova, D. G. Fraser, N. N. Vorontsov, R. S. Hoffmann, and M. J. Braun. 1999. "Molecular phylogeny of the marmots (Rodentia: Scuridae): Tests of evolutionary and biogeographic hypotheses." *Systematic Biology* 48:715–734.

Storz, J. F., and H. E. Hoekstra. 2007. "The study of adaptation and speciation in the genomic era." *Journal of Mammalogy* 88: 1–4.

Storz, J. F., A. M. Runck, S. J. Sabatino, J. K. Kelly, N. Ferrand, H. Moriyama, et al. 2009. "Evolutionary and functional insights into the mechanism underlying high-altitude adaptation of deer mouse hemoglobin." *Proceedings of the National Academy of Sciences, USA* 106 (34): 14450–14455.

Suchan, T., C. Pitteloud, N. S. Gerasimova, A. Kostikova, S. Schmid, N. Arrigo, et al. 2016. "Hybridization capture using RAD probes (hyRAD), a new tool for performing genomic analyses on collection specimens." *PLoS ONE* 11: e0150651.

Sudmant, P. H., M. S. Alexis, and C. B. Burge. 2015. "Meta-analysis of RNA-seq expression data across species, tissues and studies." *Genome Biology* 16: 287. https://doi.org/10.1186/s13059-015-0853-4.

Thompson, F. J., M. A. Cant, H. H. Marshal, E. I. K. Vitikainen, J. L. Sanderson, H. J. Nichols, et al. 2017. "Explaining negative kin discrimination in a cooperative mammal society." *Proceedings of the National Academy of Sciences, USA* 114: 5207–5212.

van der Merwe, J., E. C. Hellgren, and E. M. Schauber. 2016. "Variation in metapopulation dynamics of a wetland mammal: The effect of hydrology." *Ecosphere* 7 (3): e01275. https://doi.org/10.1002/ecs2.1275.

Vega, R., E. Vázquez-Domínguez, T. A. White, D. Valenzuela-Galvá, and J. B. Searle. 2017. "Population genomics applications for conservation: The case of the tropical dry forest dweller *Peromyscus melanophrys*." *Conservation Genetics* 18: 313–326.

Velotta, J. P., J. Jones, C. J. Wolf, and Z. A. Cheviron. 2016. "Transcriptomic plasticity in brown adipose tissue contributes to an enhanced capacity for nonshivering thermogenesis in deer mice." *Molecular Ecology* 25: 2870–2886.

Vitalis, R., K. Dawson, P. Boursot, and K. Belkhir. 2003. "DetSel 1.0: A computer program to detect markers responding to selection." *Journal of Heredity* 94: 429–431.

Wall, J. D., S. A. Schlebusch, S. C. Alberts, L. A. Cox, N. Snyder-Mackler, K. A. Nevonen, et al. 2016. "Genomewide ancestry and divergence patterns from low-coverage sequencing data reveal a complex history of admixture in wild baboons." *Molecular Ecology* 25 (14): 3469–3483.

Wang, J., E. Santiago, and A. Caballero. 2016. "Prediction and estimation of effective population size." *Heredity* 117: 193–206.

Waples, R. S. 2015. "Testing for Hardy-Weinberg proportions: Have we lost the plot?" *Journal of Heredity* 106: 1–19.

Warr, A., C. Robert, D. Hume, A. Archibald, N. Deeb, and M. Watson. 201 5. "Exome sequencing: Current and future perspectives." *G3* 5: 1543–1550. https://doi.org/10.1534/g3.115.018564.

ecology prompted entry into small mammal research. Today, his research focuses on wildlife within longleaf pine forests and the relationship between forest management, predators, prey, and the ecology of fear. Since 2003, he has served as a principal investigator on a long-term mesopredator exclusion study that has resulted in over 57,000 small mammal captures. Additionally, students in his lab have radio-tracked more than 200 small mammals. He has published four book chapters, 15 technical reports, and over 140 scientific articles. Mike lives in Bainbridge, Georgia, with his wife, Greta, and son, Eric, and their bloodhound, Jed.

James D. Austin is a professor at the University of Florida in Gainesville, Florida. Originally from Canada, he completed his PhD at Queen's University in Kingston, Ontario, and subsequently held an Natural Sciences and Engineering Research Council of Canada (NSERC) postdoctoral fellowship at Cornell University. He joined the University of Florida in 2006 and was promoted to professor in 2020. His research program currently emphasizes genomic approaches to understanding local and global patterns of neutral and adaptive variation, as well as application of evolutionary information in conservation and management of small mammals and some other, less interesting, vertebrates.

Peter J. Taylor was born in Zimbabwe and is a full professor at the University of the Free State (Zoology Department and Afromontane Research Unit) in South Africa (since January 2021). He previously headed the South African Research Chair on Biodiversity Value and Change in the Vhembe Biosphere Reserve at the University of Venda (2013-2020) (funded by the National Research Foundation and Department of Science and Innovation). He was curator of mammals and acting director at the Durban Natural Science Museum for 21 years (1989-2010). In 1994, he cofounded the Bat Interest Group of KwaZulu-Natal. His academic interests include taxonomy, conservation, ecosystem services, and ecologically based pest management, with special focus on bats and rodents. He has published four scientific books, a children's novel, over 40 popular science articles, and over 180 scientific articles.

Index

Figures and tables are indicated by f and t following page numbers, respectively.

urban gradient, 335; savanna, 39, 232; selection of, 253; tropical, 39–40, 232; urban, 44, 205, 331

hairs: collection of, 19, 22, 87, 108, 108f, 190, 284; color or shape marking via, 107–108, 109t

hair tubes, 7t, 14

hairy-nosed wombat (*Lasiorhinus krefftii*), 20

hairy-tailed bolo mouse (*Necromys lasiurus*), as an invasive species, 258

hamsters. *See* individual species of

handling techniques: bags for, 65f, 101, 126; and bag tests, 255, 255f, 257; impacts from, 190; for marking, 98, 107, 109t; permits for, 89; precautions with, 3, 64–66, 264; for scats, 285; for swabs, 85–86

Hardy-Weinberg equilibrium, 168, 339, 340

hares: in general, 1, 40; traps for, 36t

harvest mouse (*Micromys minutus*), 11

hedgehogs (Erinaceinae), 332t

helminths (parasitic worms), 73

heteromyid rodents, activity patterns of, 260

heterozygosity, 339–340, 345

Hill Numbers, 208f, 211

hispid cotton rat. *See* cotton rat

hole-board apparatus, 236f, 256–258

home-range estimates: fixed kernel, 162–164, 163f; in general, 153, 154, 165; minimum convex polygons (MCP), 162–164, 163f; utilization distribution (UD), 148, 164

homology versus homoplasy, 294–295, 326

house mouse (*Mus domesticus*; *Mus musculus*; *Mus musculus domesticus*): chew cards and, 12; ear tags and, 102t; in general, 56, 312; home ranges of, 152, 153; as pests, 240, 324; stomach contents of, 231; tracking of, 158

Huggins approach, 184, 187

hybridization, 168, 326, 331, 332t, 344

Hylomyscus, 49

hyrax (*Dendrohyrax validus*), 179

Hystricidae, 308, 314

identification of individuals, methods for: ear biopsy, 106–107; ear tagging, 88, 98–102, 106; freeze branding, 104–105; hair clipping, 107–108; necklaces, 103; PIT tags, 101–102; tattoos, 104; temporary markers, 107; toe clipping, 87, 105–106, 151

identification of species, methods for: dietary remains, 231, 233; direct observation, 3; DNA, 19–20, 22, 237, 241t, 322–323; hair, 14; images, 16–18; length,

126; molar shape; 273; museum-based, 294; predator-mediated, 273, 279–280, 282–283, 285; skulls, 45; stomach contents, 231–232; taxonomic keys, 118

independence assumptions, 137, 164–165, 180

Indian giant flying squirrel (*Petaurista philippensis*), 230

indices convergence estimator (ICE), 206

inhalants, 68–70

institutional animal care and use committees (IACUCs), 273

intrapopulation variation, 302–303

invasive species: in general, 16, 63; management of, 204; scat analysis of, 331

invertebrates: barcoding information on, 239t; communities of, influence of small mammals on, 203; eDNA studies of, 20; reference collections of, 232

Jaccard Index, 213, 217

jackknife estimator, 183, 206

jerboas (Dipodidae), diagnostic characteristics of, 308–309

Jolly-Seber (JS) population model, 186

kangaroo rats (*Dipodomys*: Heteromyidae), 180, 310. *See also* individual species of

Keen's mouse (*Peromyscus keeni*), 184

kinship, 169, 332t

labels, 17, 67–68, 76

laboratory mouse (*Mus musculus domesticus*), 83. *See also* house mouse

Lagomorpha, 1

lagomorphs: in general, 3, 8, 332t; sampling of, 179, 184; scat surveys of, 12; traps for, 39

laminate-toothed rats (Otomyini), 310

lancets, 81–83

landmarks: anatomical, 298, 300; GM, 308–309, 311–313; homologous, 300; morphometric, 295, 302–304, 306–308

landscape of fear, 265

Leimacomys, 314n*

lemmings: crania of, 311; species associations of, 312f. *See also* individual species of

Lemniscomys rosalia, 210f

length measurements, 116, 117, 126–128, 297, 298f, 299t, 308–310

leporids, 286

linkage disequilibrium, 168, 339, 345

liquid nitrogen, 72, 81, 104

Lincoln-Petersen estimator, 182–183

Longmire's buffer, 78, 80

long-read platforms, 330–331